RECENT VOLUMES

23 **CITIES IN THE 21st CENTURY**
Edited by Gary Gappert and Richard V. Knight

24 **THE GREAT HOUSING EXPERIMENT**
Edited by Joseph Friedman and Daniel H. Weinberg

25 **CITIES AND SICKNESS: Health Care in Urban America**
Edited by Ann Lennarson Greer and Scott Greer

26 **CITIES IN TRANSFORMATION: Class, Capital, and the State**
Edited by Michael Peter Smith

27 **URBAN ECONOMIC DEVELOPMENT**
Edited by Richard D. Bingham and John P. Blair

28 **HIGH TECHNOLOGY, SPACE, AND SOCIETY**
Edited by Manuel Castells

29 **URBAN ETHNICITY IN THE UNITED STATES: New Immigrants and Old Minorities**
Edited by Lionel Maldonado and Joan Moore

30 **CITIES IN STRESS: A New Look at the Urban Crisis**
Edited by M. Gottdiener

31 **THE FUTURE OF WINTER CITIES**
Edited by Gary Gappert

32 **DIVIDED NEIGHBORHOODS: Changing Patterns of Racial Segregation**
Edited by Gary A. Tobin

33 **PUBLIC INFRASTRUCTURE PLANNING AND MANAGEMENT**
Edited by Jay M. Stein

34 **ECONOMIC RESTRUCTURING AND POLITICAL RESPONSE**
Edited by Robert A. Beauregard

35 **CITIES IN A GLOBAL SOCIETY**
Edited by Richard V. Knight and Gary Gappert

36 **GOVERNMENT AND HOUSING: Developments in Seven Countries**
Edited by Willem van Vliet and Jan van Weesep

37 **LEADERSHIP AND URBAN REGENERATION: Cities in North America and Europe**
Edited by Dennis Judd and Michael Parkinson

38 **BIG CITY POLITICS IN TRANSITION**
Edited by H. V. Savitch and John Clayton Thomas

39 **URBAN LIFE IN TRANSITION**
Edited by M. Gottdiener and Chris G. Pickvance

40 **THE PENTAGON AND THE CITIES**
Edited by Andrew Kirby

41 **MOBILIZING THE COMMUNITY: Local Politics in the Era of the Global City**
Edited by Robert Fisher and Joseph Kling

42 **GENDER IN URBAN RESEARCH**
Edited by Judith A. Garber and Robyne S. Turner

43 **BUILDING THE PUBLIC CITY: The Politics, Governance, and Finance of Public Infrastructure**
Edited by David C. Perry

44 **NORTH AMERICAN CITIES AND THE GLOBAL ECONOMY: Challenges and Opportunities**
Edited by Peter Karl Kresl and Gary Gappert

NORTH AMERICAN CITIES AND THE GLOBAL ECONOMY

◆

Challenges and Opportunities

edited by

PETER KARL KRESL
GARY GAPPERT

URBAN AFFAIRS ANNUAL REVIEW
44

SAGE Publications
International Educational and Professional Publisher
Thousand Oaks London New Delhi

For information address:

SAGE Publications, Inc.
2455 Teller Road
Thousand Oaks, California 91320
E-mail: order@sagepub.com

SAGE Publications Ltd.
6 Bonhill Street
London EC2A 4PU
United Kingdom

SAGE Publications India Pvt. Ltd.
M-32 Market
Greater Kailash I
New Delhi 110 048 India

Printed in the United States of America

Library of Congress Cataloging-in-Publication Data

ISSN 0083-4688
ISBN 8039-7094-3 (hardcover)
ISBN 8039-7095-1 (paper)

This book is printed on acid-free paper.

95 96 97 98 99 10 9 8 7 6 5 4 3 2 1

Sage Production Editor: Diana E. Axelsen

Contents

Preface

Most of the chapters in this book were initially papers presented at the second conference of the New International Cities Era (NICE), a group of academic researchers who share an interest in the evolving role of cities in the international economy. Both conferences were held at Brigham Young University and at Sundance in Provo, Utah. The common issue of the first gathering was the study of the experience of individual cities, primarily in North America. The papers were edited by Earl Fry, Lee Radebaugh, and Panayotis Soldatos and were published in 1989 as *The New International Cities Era: The Global Activities of North American Municipal Governments.* For the second NICE conference (October, 1993), the focus of the research was changed to more general and theoretical aspects of the internationalization of urban economies. In addition to the papers presented at the second conference, works by John Kincaid and Arie Shachar are included to cover gaps in the coverage of the field. We also asked Wilbur Thompson to write an introduction to the volume. Gary Gappert rewrote his chapter to serve as a summary and a speculation of future directions.

Peter Karl Kresl
Gary Gappert

Introduction: Urban Economics in the Global Age

WILBUR R. THOMPSON

The global society is quite rightly depicted in these pages as consumerism writ large. Beyond this, the case is made that the global marketplace is homogenizing world culture. A recent television program followed the Avon ladies paddling up the Amazon, selling global cosmetics to natives without shoes. Kids everywhere are spending all their money on minor variations on world-class gym shoes. I add here, to the many examples in this book, the worldwide spread of sports. Europe now plays basketball and is looking at football; Japan plays baseball, and the United States is making a run at the universal sport, soccer.

Few need to be reminded, especially the readers of this book, that the motives that drive humans are varied and complex: self-expression, creativity, fame, and power. Therefore, alongside the homogenizing power of global consumerism, other forms of differentiation will flourish. Urban scholars and other social scientists argued in the 1960s that social stratification was increasingly taking the form of occupational specialization. But even here are elements of global homogenization. Transportation planners in one country know better and can talk more easily with their counterparts in other countries than with the land planners next door, unfortunately. But to the extent that work is still treated, conceptually and organizationally, as a cost to be minimized—as a means, not an end—consumerism (better *consumer sovereignty*) commands the field.

Consumers, Producers, and the Distribution of Income

Champions of global competition and free trade are quick to note that consumers benefit greatly by being able to shop in a worldwide market for the best product or the one at the lowest price. But they are less quick to point out that for producers, both businesses and workers, worldwide competition is a mixed bag, with about as many losers as winners. Although the word *consumerism* does succinctly describe the motivation and behavior of individuals, the older term *consumer sovereignty* captures more of the essence of the underlying, governing mechanism, expressing better the basic philosophy of the self-regulating market system that is displacing competing economic systems.

Free trade has, in fact, much in common with those other equally hallowed words: invention, innovation, and entrepreneurship. They all focus on the larger size of the pie and have little to say on the size of the shares. The dark side of these bright facets of economic progress is that they all entail sharp changes and abrupt displacements, foreshadowing at least frictional, usually structural, and perhaps chronic unemployment. All of the writers in this volume seem to agree that a central element in the new global competition is the force-feeding of invention and innovation (change) into the system.

The real debate on the North American Free Trade Agreement (NAFTA) and the General Agreement on Tariffs and Trade (GATT) should have been not on whether freer trade will cost jobs but instead on how the political-economic system can and will arrange an orderly reemployment of those sure to be displaced—and who will pay. The benefits of global competition will usually be much too great to resist; the challenge for the foreseeable future will be how to soften the blow to the losers. The process of change could be slowed to allow time for adjustment with, say, gently sagging tariffs. The adjustment of producers and workers could be eased and speeded with low-cost capital and retraining. Income could be redistributed from the winners to the losers.

Therefore, although the case made for the rise of global cities is convincing, the parallel case made for the decline of nations is more debatable. Producers see the national authorities as their last line of defense, with tariffs, quotas, and other impediments to free markets as the weapons of war. The sovereignty of consumers will be repeatedly challenged by producers, with the national government their advocate. The state and local governments, however, hold the key to workable compromises, because they play the critical roles in education, retraining, and other preparation for new and different work.

Environmentalism in counterpoint. Perhaps other goals, such as environmentalism, will arise to modify rampant consumerism. Even young people, who are perhaps the most ebullient consumers, may, as they mature and gain the self-confidence needed to assert their individualism, show changing priorities in their life cycles. Many young adults are the most enthusiastic protectors of the environment, but this too can change with aging.

Will an increasingly intense global competition improve or degrade the environment? Competition will probably improve the environment in places to which mobile, educated, demanding workers must be attracted, such as advanced service centers and most tourist destinations. But places of routine operations, hard-pressed to minimize costs of production, will not feel that they can afford to trade amenities or even health and safety for jobs. Global competition will probably produce greater divergence in natural environments.

Many guardians of the world's resources and natural environment claim that overconsumption results more from the excesses of rich nations than from population increases in the poor ones. One can understand, then, how environmentalists would have mixed feelings when asked to cheer a worldwide flood of goods and services flowing from an ever more productive global economy. The rise of the global economy and freer trade seems not so symbiotic with global environmental protection. How can the lot of poor nations be raised with only moderate increases in total world output and effluent except by reducing consumption in rich nations, ones in which the younger generation does not feel so rich? And who indeed is in charge here? At every turn of these pages, the distribution of income seems to be the central issue.

To carry the argument to the local level, the point is raised that places that address environmental issues first will enhance their long-term competitive advantage. True, but the fruits of victory may be a little bitter if more local growth leads to renewed pressure on the local environment. It is hard to write a scenario in which city *growth* does not clash with the protection of the environment. But, as argued subsequently, *development* could continue if the global city judiciously discarded old pieces of growth as it added new ones.

Governance: The Changing Balance in the Federal System

The prevailing opinion in this book is that the global city must be self-governing to act with world-class decisiveness and dispatch. But most central cities in the United States contain less than one half of the population of the local economy (local labor market), and many have less than one quarter of the local population and labor force, for example, Atlanta (13.9%), Boston (23.5%), and Pittsburgh (18.0%). The central city cannot easily plan for full employment and, with more than its share of the poor, will be hard-pressed to service the poor. Again, central cities without command of the full commuting radius will struggle to plan and manage land use patterns and transportation systems.

Besides, most large cities are already multinodal, posing the threat of destructive competition for fiscal advantage, rather like feudal barons in office towers. Moreover, the geometry of the physical growth of those cities on waterfronts (most of them) almost ensures that the old center will be off-center. Finally, about one third of the largest central cities continued to lose population between 1980 and 1990—in absolute number, not just in share of the metropolitan area, that is, they were becoming even less central to all-inclusive local governance.

One can easily sketch two scenarios. In the first, the central city "island" succeeds in its global networking, leading to higher land values in the recycled core area, with only the highest value-added activities able to pay the rent. Students of government should feel uneasy that competitive success will almost surely take the form of rising land rents and housing prices and that central cities will become home largely for winners of the world games. In the second scenario, the central city languishes for years, and the new offices settle largely in suburban centers out of reach. The central city can become a strong and responsible government only if it includes all income classes and most workplaces and homes—both the origins and the destinations of movement. Only then is it likely to supplant the national government as the dominant player in economic development.

In addition, disequilibrating processes tend to dominate in central city-suburban relationships. Poor areas are forced to raise taxes or cut services, succeeding only in driving out the more mobile affluent households. Rich areas can win the competition hands down by keeping taxes attractively low and services comparatively good. Only some overarching public authority, some form of metropolitan government, can show the strong, prompt response needed to dampen the destructive side of "voting with one's feet."

All in all, the new global cities, founded on advanced services, seem destined to perform much like the regional trade and service cities that preceded them, creating more rich and poor. This stands in sharp contrast to manufacturing areas that were more egalitarian. If a central city island of corporate headquarters, no matter how powerful and vital in the global network, cannot ensure either urban efficiency or distributive equity, how can it be truly self-governing? This leads back to 1960 and the debate over metropolitan area government, without much better prospects now than then. Where is the municipal revenue sharing that would balance tax base and service need and head off the cutthroat competition that distorts efficient land use patterns?

Role of the state in economic development planning. Many of the contributors to this volume see cities joining forces to lobby for government programs and exchanging information and experiences with others to achieve "best practice" programs and to combat common problems. Maybe the representatives of cities and counties should also be urged to trade with each other in the halls of the state capitol for selective state support: "We will support your bid for the new airport if you support our claim for the new center of medical research." New city-county intelligence agencies (CIAs), armed with better databases, would see better what their local economy could give up most readily to get more of what it most needs. Although the authors in this book discuss the role of the new global city vis-à-vis the federal government in economic development planning, they leave open for the next round of writers the role of the state. In the absence of a much greater national government interest and aptitude in local economic development, the state could fill a political vacuum until metropolitan governments are formed or until central cities show that they can dominate or lead the surrounding ring.

The state holds a lot of high cards in the game, apart from how well they are played: higher education, transportation systems, regulation of utilities, and environmental protection. In fact, in Canada, the provincial governments hold almost all the cards in urban affairs. Although the state has few free funds for local economic development per se, just spending better the money they do have in hand for higher education, vocational education, and transportation can be a powerful tool in local economic development. Nothing is ever simple: The location of many large cities on state borders (New York, Philadelphia, Chicago, Cincinnati, Louisville,

Memphis, St. Louis, Kansas City, and Portland) will require interstate compacts to carry out inclusive strategies.

Federal versus local industrial planning. About 40 years ago, the conventional wisdom in the literature on public finance envisioned the federal government as responsible for economic stabilization (through monetary and fiscal policy) and for the redistribution of income (through the federal income tax) and saw it sharing with lower governments the responsibility for the provision of public goods and services. If foreign trade policy now hobbles monetary policy, if politics constrains fiscal policy, and if state and local governments do a poor job in delivering education, the federal government will be hard put to ensure full employment, especially if the chief foreign competitors are doing a better job in teaching science, math, and expository writing.

Public finance economists have never fully appreciated the key role of local government in redistributing income, expressed largely through schools and community planning. By concentrating on the progressive personal income tax, they have almost ignored the role of education and community environment in the redistribution of *opportunity* and, ultimately, income. Preparing the local labor force for employment is arguably the most permanent form of income redistribution: "Teach a man to fish . . ." The new central city global islands may network well but seem ill-prepared to assume the lead role in the redistribution of opportunity and income through public services.

But there is a new (tough) kid on the block: local economic development. These chapters raise the important issue of whether cities may come to do more, perhaps most, of the industrial planning, with local education programs linked closely to local industrial and occupational profiles and targets. The case for metropolitan government is strengthened by adding industrial-occupational planning to the earlier arguments for economies of scale (e.g., water and sewage plants) and the redistribution of tax base. Perhaps the fierce competition of global markets and a worldwide labor pool that threatens chronic, structural unemployment will tip the political balance. It would give a strange, new twist to industrial planning if local government were entrusted with the much disparaged job of picking industrial winners and losers, even if the private sector—the local chamber of commerce—were to do the picking with their bottom line on the line. What should be the new rules for the time-limited protection of "infant industries," and who would be empowered to practice "industrial euthanasia"?

The Economic Base of Cities: Changes in Type or Just in Degree?

The arena is bigger, but the name of the game is still the same: export or die. The concept of a system of cities is long-standing. The difference is that the traditional concept emphasized a national system of regional trade and service centers with surrounding rings of satellites and rural areas, whereas the global city is seen more as part of a worldwide network of more or less similar-size, competing cities. But because hierarchies exist even among these global cities (e.g., New York dominates Atlanta in corporate management), central place theory is alive and well, although living in a bigger house.

Filtering down of industries and occupations in the network (system) of cities. The new global society is unlikely to change the longtime process whereby the largest, most sophisticated places must innovate to replace old activities that are sliding down the learning curve toward smaller places that can perform the ever simpler work as well and at lower cost. The essence of economic development (compared with raw growth) lies in the spinning off of older, lower functions to make room for newer, higher ones. But the cost of this new big-city version of "lean and mean" is a frequent, severe, and unending displacement of workers, combining elements of both frictional and structural unemployment.

Although this filtering sometimes takes the form of whole industries (textiles moving from New England to the Carolinas), it often is confined to separable, simpler functions within industries (automobile assembly leaving but managerial and research functions remaining in the Detroit area). Today, I would shift the emphasis in job filtering from industries to functions and concentrate on identifying the new occupations that carry the seeds of economic renewal. In health and medicine, one no longer has to travel to the Cleveland Clinic or to the Houston Medical Center to get first-class heart bypass surgery. To remain a world-class medical center today, a place must achieve distinction in the more risky heart transplants. In finance, the cutting edge could be in new types of derivative securities or new applications, such as in the extraction of natural resources.

Services can be just as volatile in the new global economy as manufacturing has been in the past. Note the effect of terrorism on tourism. As marketing geography changes and as multinational corporations pass through mergers and acquisitions, headquarters can be moved as easily as factories.

Stability through diversification or through adaptability. Playing in this new world-class league will surely call for greater industrial and occupational specialization to achieve comparative advantage. Where will this leave the goal of industrial diversification, always near the top on every development strategist's laundry list? Growth stability (in contrast to cyclical stability) may have to be achieved less by portfolio diversification and more by adaptability. *Basic* education may be the leading new index of resilience, whereas *specialized* education has mixed effects. A place loaded with electrical engineers has many options of new industries with which to replace a faltering one, such as military aircraft. But places built on the skills of mining engineers look in vain for a replacement industry (a new use of that skill) when the mine plays out or the ore price plummets.

These chapters argue, however, that a strong local culture can breed civic loyalty that reduces labor mobility and stabilizes the community by prompting talent to hang in there through the tough times—those periodic transitions in the export base. Conversely, the multinational corporation increases mobility and undermines that civic loyalty. The trends here and the changing balance surely vary from place to place and from job to job.

Targeting sounds good: What should we target? It is unclear whether these ever larger global cities are indeed developing a broader range of specializations—really becoming more diversified as some claim—or are just becoming larger agglomerations of linked industries: vertically integrated complexes of headquarters, associated business services, and research labs all founded on a narrow base of just a few final products. In fact, economic development consultants exhort localities to target clusters of linked industries for maximum effect on the next page after the one on which they call for more industrial diversification; see my article in a recent issue of *Economic Development Review* (Thompson, 1994).

The same issue of that journal was dominated by articles in which authors advocate concentrating on industries producing a relatively high value added per worker, that is, industries that employ workers with above average education and skills. Presumably, everyone in town is above average in economic potential, if not in present capability. Such narrow targeting neither ensures domestic tranquility nor promotes the general welfare. Only I argued for *industrial and occupational* targeting aimed at the weak spots in the local labor market, as well as opportunities for skill enhancement.

Central cities can responsibly pursue high-value-added industries and occupations only if they also plan for low-skill workers. For example, central city development planners could try to guide routine manufacturing firms toward the smaller cities in its hinterland to

1. find the least-cost location—lower land rents and living costs (lower money wages for the same real wage);
2. slow growth of the central city to avoid congestion, reduce concentrations of effluent, and dampen the inevitable price rise; and
3. focus central city development on high professional and managerial services, rendered in significant measure to that industrial hinterland, variations on the advanced services exported to its global markets—finance, accounting, engineering, and computer services.

But typically, enthusiastic city economic development managers try to bring everything that moves into town. It is easier to keep score with raw numbers of jobs and to pile up political points. True, market mechanisms will *eventually* force least-cost locations but at the economic cost of unanticipated relocations that accelerate the depreciation of linked private and public investments and at the social cost of abruptly disrupted lives.

Toward typologies that work. Everyone seems to agree on three world cities: New York, London, and Tokyo. But this set of full-service global cities is much too small to encompass this wide-ranging subject matter and too small to attract the broad interest the subject merits. Writers in this field are well into the work of defining a few broad classes of international cities, that is, cities that interact significantly with other parts of the world. It is, in fact, easy—even fun—to name different limited-service global cities: Washington, DC (government); Mecca, Jerusalem, and Salt Lake City (religion); Acapulco, Honolulu, and Rio de Janeiro (tourism); and Detroit (automotive technology).

Ideally, the taxonomy would connect different export sectors to various profiles of economic performance and diverse paths of economic development. For example, a small, global "niche city" (Peoria), exporting a producer durable good (construction equipment made by Caterpillar), might experience even greater instability than it had grown accustomed to when selling mainly in the domestic market, because it is now burdened with intense Japanese competition as well as a harsh business cycle. A small linked-exporter, such as a marginal supplier of valves and fittings for a Houston exporter of oil drilling machinery, would be especially subject to boom and bust ("tapered vertical integration"). Turning to services and the long run, some authors in this book suggest that places heavily engaged in world tourism often find that their global networking does not lead in turn to higher functions and occupations—economic development. The global city's new economist should become a good industry analyst.

The large manufacturing cities of the Midwest pose the most difficult problems of transition and transformation. Their difficulties of transition are probably overstated by pessimists, who seem less impressed than I am by the "ratchet effect"—the grace period provided by large size. Note that Detroit-Ann Arbor, Cleveland-Akron, Pittsburgh, and St. Louis all have prestigious universities, elaborate research parks, and impressive museums. On the other hand, the time it will take to transform a manufacturing center into a true city seems always to be underestimated by the optimists who seem less impressed than I am by the agonizingly slow aging and replacement of housing and public capital (in sharp contrast to the rapid technical obsolescence of business capital). The weak downtowns of manufacturing areas may be a drag on the development of an advanced service export sector, but again their multinodal form may prove to be just as efficient an intra-urban stage on which to play a global role.

Toward an operational occupational-functional classification. As an alternative to classifying cities by industry, I have suggested that the development path of a city can be seen as some combination of five distinct functions: entrepreneurship, central management, research (science and engineering), precision production, and routine operations (Thompson & Thompson, 1987). Although these five paths have been quantified, experimentally, for all U.S. metropolitan areas, the global system of cities seemed a long reach a decade ago, given the limitations of the data. It would be ideal to document the reputed entrepreneurial flair of Monterrey, Mexico, and compare it with the job shops of Taiwan. Can the profile of a German metalworking city be quantified and used as an international standard of precision production work?

Even routine operations can create an international production site, but the pressures that accompany urban growth on the cheap tend to inhibit development. The chapters here detail the case of border cities (e.g., with *maquiladoras*) that enjoy rising per capita income without finding many new, higher rungs on their occupational ladders. These places are building up truck traffic, industrial pollution, and cardboard houses lacking water and sewer, without ever catching up in the needed tax base. A number of these chapters explore the difficulty of coordinating city planning across national boundaries: different languages, customs, and political and social systems and official turf battles about problems over which no single government has jurisdiction.

Even these low-skill production centers could, with good basic and vocational education and on-the-job training programs, serve a national and even international function as "farm clubs" for the world-class league. Perhaps a class of internationally linked places will never develop, much like rural areas in developed countries. Places built on low-wage, low-skill operations are at risk because they tend to flourish where there are few amenities and could be abruptly abandoned when they wear out, mimicking the cut-and-run patterns in logging and the mining towns nestled among hills of overburden.

Critics charge that many a global city may be little more than an impermanent cluster of multinational corporations. Whether a city is more than this depends on whether that place remains in the global network as the current set of corporate tenants changes. Would Detroit still be a global city if its automobile companies contracted even more or moved away—first steadied and then transformed by its sheer size and technological base? Perhaps global Seattle would survive the loss of Boeing because of geography. Clearly, there are global sites on which lucky cities sit.

■ Integrating Regional and Urban Economics

As the new frontier in regional and urban economics, I would choose the integration of what is usually treated as separate subject matter. How do the networking global cities differ from each other and differ from traditional regional trade and service cities in the way that they affect the form and functioning of intra-urban space? And how do various intra-urban patterns act in turn to strengthen or weaken different export bases?

Different intra-urban platforms for different export bases. Almost 30 years ago, I tried to link the export base of a place to its internal performance and intra-urban form. The large manufacturing cities of the 1960s were seen as leaving a legacy of median family incomes that were above average, more equally distributed, and more unstably generated. Specialization in manufacturing seemed also to leave a legacy of weak central business districts, poor and depopulating central cities, and auto-dominated transportation systems struggling to handle cross-commuting and cross-hauling traffic between many outlying centers. Some of this has changed through the years: For example, manufacturing cities now generate only an average income and the relatively equal income distribution is now confined to the smaller set of workers who remain employed.

Moreover, it is not all clear that the downtowns of the New Age, global, high-service cities will dominate the skyline with the largest number of office towers and will validate hub-oriented transportation systems. The new taxonomy of intra-urban patterns could differentiate in the relative strength of downtowns and in the type and degree of multinodality, distinguishing between advanced service cities that feature world headquarters of U.S. corporations (Chicago), ones with mostly regional headquarters (Atlanta), and ones that appear especially attractive as sites for foreign branch offices (often simply because the downtown is more familiar). Although the different export bases of these new global cities may rest best on different intra-urban platforms, the conditioned response of city planners seems to be to try to stuff divergent industrial and occupational-functional mixes into a small, familiar set of either classic or ideal urban forms—or make a vain attempt to rescue old investments.

Transportation and land use planning. To control the intra-urban land use and transportation pattern, the central city must command at least the full urbanized area and, preferably, the next ring of open space out to the edge of the job-commuting radius. The competitive city should offer a wide choice in transportation options (but at near full-cost prices, including social costs) and should also shorten trip distances to accommodate more trips and more interaction with less movement. Ironically, as we look to buy products from afar, we cannot seem to bring the home and the workplace closer together.

Unfortunately, land use and transportation planners do not have a history of working well together, largely because their objectives as narrow specialists are often seen as competitive and contradictory rather than as complementary and supportive. A major part of the solution to the urban transportation problems of congestion and huge fixed investments lies in bringing origins and destinations closer together in space—reducing movement and diminishing the transportation planners' role. In turnabout, the substitution of transit for the automobile will come in significant measure only through transportation corridors, which are hardly an aesthetic favorite of land planners and landscape architects.

On the subject of transportation corridors, I have urged that cities should not promise more in public transportation service than they can deliver (Thompson, 1973). Too often, local public officials and managers have promised, implicitly, that one could choose to live anywhere in the area and be served with good public transit—a promise the public sector could not keep as long as most of the population opted for automobility. Rather than spending their transit money evenly across the whole area, cities should experiment with "thick subsidies for thin systems" to implant a stronger structure into those amorphous transportation systems and land use patterns. The transit-preferring and transit-dependent population would be expected to move near the guaranteed, high-quality, skeletal, low-cost system.

The interaction between intra-urban form and the city export function is clearly seen when we recall that only large cities can have the extensive number of flights required to create the good air schedules critical to being a global player. But a huge population creates high density and congestion on the ground, unless cities learn how to avoid being bigger by selectively filtering out lesser functions as they add higher ones. The "unseen hand" may do this automatically, if sluggishly, with higher land rents and traffic congestion. As routine manufacturing leaves the region and retailing leaves downtown, congestion becomes an intra-urban "solution" that exacerbates the income distribution problem under political fragmentation and that undermines that "self-governing city."

Depopulation and repopulation: Recycling whole cities. With the era of urban population growth ending—birth rates are falling and rural areas have about emptied out—cities can turn from accommodating growth (e.g., force-feeding housing into packed cities) to planning the rearrangement of the furniture. Even absolute depopulation of the central city can be as much an opportunity as a problem if it is seen as a chance to recycle the whole city. The next trip through the building of the city could feature mixed-age housing, mixed at finer grain. The concentric zone form was efficient or at least unavoidable the first time through, but it led inevitably, through inexorable aging, to those ugly "holes in the doughnut." Balancing the age-mix of housing is critical for bringing the home and the workplace closer together to reduce superfluous movement, for balancing tax base and public service needs under political fragmentation, and for promoting social mixing—for example, James B. Conant's (1961) "comprehensive high school . . . as an instrument of democracy" (p. 4).

Because the large manufacturing cities have suffered the heaviest depopulation (Detroit, St. Louis, and Cleveland), these limited-function global cities will offer both the greatest problems and greatest possibilities of land use change in their second lives. And so the time has come to reexamine alternative depopulation strategies for older residential areas, such as "clustering and clearing" in old-age residential areas versus random "thinning." Population densities are falling everywhere as higher incomes are used to buy privacy, but the trade-off is that public service efficiency would usually be better served by increasing densities in some areas and emptying out others, however scary the politics.

Toward a public price policy to discipline city dwellers and local officials. Cities are shaped largely by free market forces, but these forces do not include all the social costs and social benefits. The new city economist will need to devise a set of managed prices to correct this constrained accounting. A new overseeing hand is needed to ration scarce space and to allocate public investment, but the hand should be clearly seen and guided by a clear head. A few random tolls and fees to raise revenue do not constitute a broad price policy.

Ideally, public prices would include cost-of-service-based property taxes to achieve both efficiency and equity, a shift in large part from regulation to user charges in pollution control, tolls on roads in key places at peak travel times, low or no-fare buses, and so forth. It has become almost commonplace to talk about specific user charges, but each has its own side effects and feedbacks, such that a systemic approach is as central to the work of the city economist as comprehensive planning is to the holistic city planner.

For example, the no-fare bus acts, in itself, to increase city deficits and subsidizes movement and sprawl, unless financed with automobile tolls. Again, as part of the "social contract" between the city and those who agree to move into the new transit corridors, argued above, the city economist would need to include in the new comprehensive price policy some measure designed to deflate land prices near the transit stops so transit-dependent poor people could afford to live there. For example, the taxation of capital gains on nearby land, used to finance low-income housing along these routes of high public purpose, would correct for unintended side effects.

Prices not only ration scarce supplies and allocate resources but also finance production. Some of the chapters in this book distinguish sharply between the financing of a service with taxes or fees and the provision (production) of that service, arguing that the latter is often more efficient if privatized. Note also that it is politically much easier to defend user charges in private refuse collection than to charge taxpayers for a "public service."

Certainly, those who champion the free market as an efficient allocative device in the worldwide location of industry should see a natural extension of their arguments in managed prices that discipline urban form and function. If all of this seems politically unrealistic, the hope here is that fierce global competition, leading to chronic, structural unemployment, will trigger the survival instinct and force hard choices that might not have passed muster on arguments based solely on health, safety, convenience, and aesthetics.

Unfortunately, the conventional political wisdom is that the leader of the flock should talk about just two or three things at one time to avoid blurring the message. To lump together growth, housing, land use, transportation, industrial targeting, human resources training, and employment at the same time is to confuse the electorate and the legislature. But, in the urban system, a partial solution here tends to block or at least constrain a linked solution there. Maybe it is too late and the game too complex to "keep it simple."

From Concepts to Measurement: Looking Back to Look Ahead

The many provocative speculations of these globalists set the stage for the next step: the framing of testable hypotheses and extended testing. Cities are reactive in large part because they are timid in forecasting the future. One way to predict the future is to examine the past—and not just the immediate past, because it is hard to distinguish the fashionable emerging trends from temporary aberrations or random events. I am now finding many new insights into the nature of local development, both metropolitan and nonmetropolitan, through a careful examination of 50 years of census data, all neatly packaged in the five, closely comparable decennial censuses since 1950 (see *Decennial Census of Housing* and *Decennial Census of Population,* 1950-1990).

For example, there are clear, strong trends and long waves among the major determinants of the local level of income, the single best measure of economic development. Indexes that showed significant correlation in 1950 have faded by 1990, and new ones have taken their place. Even the new favorites show signs of giving way to a new generation of basic factors. For example, the percentage of local high school graduates was a leading indicator (determinant) of local median family income in 1950, but it has since been replaced by the percentage of those who are college educated.

Because today almost everyone has graduated from high school, that variable does not vary significantly anymore. But more important for the futurists among us is that with rising rates of completion of college, the census must now gather more data on postgraduate education to keep abreast of the times.

Local female labor force participation rates have risen from no correlation with local income to a position rivaling rates of college-educated persons. But it too is becoming a variable that does not vary much between places, suggesting that more advanced indexes, such as percentages of women in managerial work, are already more sensitive indicators of local economic development. Given the importance of managerial work in the global network, leadership in tapping the potential pool of female managers could indeed provide a competitive edge, especially considering the great cultural differences around the world in the role of women.

Long-term trends in the female labor force participation rate have powerful and complex effects on the global environment. Employed women boost family incomes and reduce family size by having fewer children. The net effect on total income and consumption of smaller populations with higher per capita incomes is hard to assess, and therefore the change in the pressure on world resources and on the environment becomes a moot question. Because some consider the consumption per capita in rich countries to be more the cause of world overconsumption than sheer population growth in poor countries, it is possible that changes in family lifestyles could exacerbate environmental problems.

To predict the effect of globalization on trends in urbanization and on the prospects of individual cities, urban policymakers and city planners will need to assemble much more ambitious databases than are now in hand. Perhaps the highest hurdle here is that few countries collect statistics that match those of the United States in scope, quality, span of time, and comparability. And even most U.S. urban social scientists have not seemed to notice the quiet assembly of this country's unique 50-year census database—or else the required investment in time does not fit well into the tenure and promotion calendar of scholars.

Another dimension of the global society that begs for recognition is the role of small places, even rural areas. They may never be global places, but already they are global pawns and need to learn how to cope. But, more than this, in this age of instantaneous communication, the new "lone eagles," working in the shadow of a mountain with powerful personal computers, modems, and fax machines, have both the global library and their peers close at hand. It takes less time to communicate with almost anyone, anywhere, than to commute to work in a global city. But perhaps that is the next book in the series: *Everyman's Guide to the Global Society.*

REFERENCES

Conant, J. B. (1961). *Slums and suburbs: A commentary on schools in metropolitan areas.* New York: McGraw-Hill.

Decennial census of housing. (1950-1990). Washington, DC: Bureau of the Census.

Decennial census of population. (1950-1990). Washington, DC: Bureau of the Census.

Thompson, W. R. (1973). A preface to suburban economies. In L. H. Masott & J. K. Hadden (Eds.), *The urbanization of the suburbs* (Urban Affairs Annual Review, Vol. 7, pp. 409-430). Beverly Hills, CA: Sage.

Thompson, W. R. (1994). An occupational-functional approach to targeting industry. *Economic Development Review, 12*(2), 38-44.

Thompson, W. R., & Thompson, P. R. (1987). Alternative paths to the revival of industrial cities. In G. Gappert (Ed.), *The future of winter cities* (Urban Affairs Annual Review, Vol. 31, pp. 233-250). Newbury Park, CA: Sage.

Part I

North American Cities and the Global Economy: An Overview of the Major Issues

1 North American Municipalities and Their Involvement in the Global Economy

EARL H. FRY

The United States, Canada, and Mexico each have federal systems that divide authority constitutionally between national and regional governments. In addition to 10 provinces, 81 states, 2 federal districts, and at least 5 major territories, these North American nations also have approximately 40,000 cities and towns, including 46 metropolitan areas with populations surpassing 1 million and approximately 280 cities with populations exceeding 100,000.[1] In addition, each country ranks among the most urbanized in the world, with almost three quarters of North America's residents living in urban areas and more than 40% concentrated in the 46 major metropolitan regions.

In the nearly half century since the end of World War II, two trends are evident in the development of nation-states and municipalities. First, there has been a marked proliferation in the number of nation-states, with membership in the United Nations almost quadrupling between 1945 and 1993. Second, there has been a dramatic increase in both the rate of urbanization and the population base of cities. The percentage of people living in urban areas worldwide increased from 29% in 1950 to 43% in 1990, with urbanization up from 54% to 73% in the developed countries and from 17% to 40% in the developing nations (Williams & Brunn, 1993, p. 12).[2] Moreover, by the year 2000, 35% of the global population of more than 6 billion is expected to live in communities with more than 100,000 inhabitants, up from 27% in 1980 and 6% in 1900 (Williams & Brunn, 1993, p. 2). This thrust toward urbanization has also characterized the development of American, Canadian, and Mexican societies through the latter half of the 20th century. As North American municipal leaders prepare to enter the 21st century, however, they will find that the governance of their cities will be complicated by three quite divergent contemporary trends: (a) growing global and regional interdependence that will reduce some of the decision-making latitude at the level of the nation-state; (b) increasing intermestic pressures on many national and subnational governments; and (c) serious centrifugal forces at the local level, especially in the major metropolitan areas. This chapter will first examine these divergent trends and how they are affecting municipal governance. The latter portion of the chapter will then focus on how U.S. and other North American municipalities can adapt to increasing global and regional economic interdependence, with special attention accorded to international trade, investment, and tourism activities.

Interdependence, Intermestic Policies, and Local Cleavages

Global Interdependence

In the current literature devoted to international relations, terms such as *global interdependence* and the *global village* are commonplace. In particular, economic interdependence has evolved to unprecedented levels during just the past few decades. Global trade in goods is now approaching $4 trillion per year and trade in services $1 trillion annually, with combined trade up almost 70-fold in nominal terms since 1950 (*International Herald Tribune,* 1993a, p. 4). In addition, many nations are in the process of forming regional trading blocs. Now that the Treaty on European Union has finally been ratified by all the member states, the European Union (EU), with 12 nations, 350 million people, and an almost $7 trillion combined gross domestic product (GDP) in 1994, will begin to phase in its ambitious regional economic integration and expansion program.[3] On the other side of the Atlantic, the North American Free Trade Agreement (NAFTA) went into effect in 1994 and includes the United States, Canada, and Mexico, with a combined population base of 375 million. Eventually, NAFTA might be transformed into a hemispheric free trade accord involving most North, Central, and South American countries from the Arctic Circle southward to Cape Horn. Across the Pacific Ocean, the six Association of Southeast Asian Nations (ASEAN), which includes Brunei, Indonesia, Malaysia, the Philippines, Singapore, and Thailand, with a combined population base exceeding 325 million, began during 1993 to phase in the

Asian Free Trade Area (AFTA), a trading arrangement designed to eliminate tariffs in 15 key sectors during a 15-year period. In the Southern Hemisphere, the Southern Cone Common Market (*Mercosur*), composed of Argentina, Brazil, Paraguay, and Uruguay, is also dedicated to ending most tariff and nontariff barriers by the end of 1994 in a region with 200 million consumers. At a meeting held in Cartagena, Colombia, in June 1994, these four countries jointly pledged with the 15 other largest nations in Latin America to create a continentwide free trade zone.

Proportionally, international direct investment (IDI) has increased even more rapidly in recent years than global and regional trade, catapulting 10-fold in real terms between the end of the 1970s and the end of the 1980s. Between 1983 and 1989, IDI flows increased at an annual rate of 29%, three times faster than the growth in global merchandise exports and four times faster than the growth in individual national economies.

Despite the close linkages between trade and direct investment activity, IDI is often overlooked as a force promoting interdependence. Approximately 37,000 transnational corporations (TNCs), up from just 7,000 two decades ago, now control more than $2 trillion in direct investments, with investment placements increasing by $150 billion in 1992 alone. These TNCs possess about one third of all private sector productive assets around the world. The sales by TNCs outside their home country totaled $5.5 trillion in 1992, greater than total global trade in goods and services.[4] Moreover, perhaps 30% of global trade is composed of transfers between a parent firm and its subsidiaries, solidifying the linkage between trade and investment flows (Camilleri & Falk, 1992, p. 70).

Furthermore, perhaps no area so dramatically illustrates the growing vulnerability of national governments to international and transnational economic developments than the financial sector. In a vain effort to maintain the narrow range for currency fluctuations within the EU's European Rate Mechanism (ERM), the German Bundesbank expended 60 billion deutsche marks ($35 billion) and the French central bank used 300 billion francs ($50 billion) to support the franc in late July and early August of 1993 (*International Herald Tribune,* 1993b, p. 9). This was a losing cause, and an emergency meeting of the EU finance ministers resulted in the ERM being effectively shelved for the time being. In general, the evolving transnational economy is affected more and more by the flow of money, with such transactions easily dwarfing flows in goods and services. As Drucker (1990) has observed, the "monetary and fiscal policies of sovereign national governments increasingly react to events in the transnational money and capital markets rather than actively shaping them" (p. 115).

No central bank can withstand the pressure placed on its national currency if the private investment community makes a concerted effort to influence the value of that currency. Japanese government and central bank representatives recently learned this bitter lesson as they helplessly watched the yen rise in value against the U.S. dollar and other major world currencies, a development that may portend significant difficulties for Japanese merchandise exports. Only by working together, reminiscent of the famous Plaza Accord of September 1985, which was designed to lower the value of the U.S. dollar, can the major central banks have a noticeable effect on transnational currency markets. Yet even in this case, the combined financial power of the private sector is still capable of overwhelming the combined reserves of these central banks, indicating that cooperation not only must extend across national borders to other governments and government agencies but also must extend at times to major private actors in the transnational economy.

The transportation revolution has also helped more people than ever before to learn firsthand about what is transpiring outside their national boundaries. International tourism was at record levels in 1992, with 476 million international tourists spending in the range of $268 billion. In contrast, there were an estimated 25 million international tourists in 1950, about 1/20 the level of 1992 (Knight, 1989, p. 27). Furthermore, many countries have a growing dependence on foreign visitors as a vital source of revenue. For example, in each of the past several years, foreign tourism in the United States has been the leading earner of foreign exchange, exceeding revenues derived from exporting aircraft, automobiles, or agricultural products.

The Intermestic Dimension

In an effort to cope with increasing global economic interdependence, noncentral governments in both the United States and Canada have rapidly expanded their activities in the international economic system. As of 1970, only 4 of the 50 U.S. states had opened an office overseas; 20 years later, 43 states were operating 160 offices abroad, four fifths of which were located either in East Asia or Western Europe. Although Ontario has recently closed all 17 of its overseas facilities and British Columbia is shutting down several others, Canadian provincial governments still maintain about 40 offices abroad, a greater number proportionally than the U.S. states (*Globe and Mail,* 1993, p. B8).[5] Annually, 40 U.S. governors and most provincial premiers now lead at least one international mission in search of trade, investment, and tourism opportunities, and states spent $92 million on international economic programs in 1990, excluding investment incentive programs (*Governor's Weekly Bulletin,* 1990, p. 1; *Nation's Cities Weekly,* 1991b, p. 6).[6]

At the municipal level, many mayors of big cities in North America also headed international missions, and leaders of smaller cities in the United States can participate in the periodic missions sponsored by two umbrella organizations, the National League of Cities and the U.S. Conference of Mayors. In 1990, almost 1,000 U.S. communities had also teamed up with 1,850 municipalities in 96 countries through the sister city program (Zelinsky, 1990, p. 43). Increasingly, these sister city alliances are used to strengthen economic linkages between the twinned municipalities.[7] Cities generally rely on their state offices, U.S. embassies, and the more than 130 overseas offices of the U.S. Foreign and Commercial Service situated in approximately 70 countries to represent their economic interests abroad. Nevertheless, Tucson, Arizona, opened its own trade office in Taipei in 1987; the Las Vegas Convention and Visitors Authority maintains representative bureaus in Tokyo, Frankfurt, and the West Midlands in England; and several major California municipalities have joined with the state-level Division of Tourism to open representative offices in the United Kingdom and Germany (*Business America,* 1993b).[8]

A variety of factors account for the rapid increase in noncentral government involvement overseas (see Fry, 1989). Above all, the imperatives of complex global interdependence are pushing these subnational governments to be active participants at the international level. The term *intermestic policies,* which means the growing overlap of international and domestic issues, captures the essence of the new challenges facing these subnational governments in North America. In effect, the lives of local constituents are being increasingly affected by actions that occur and decisions that are rendered outside the boundaries of their own nation-state.

For example, in the realm of international direct investment, the United States ranks as the number one host nation in the world. Five million Americans now work for foreign-owned companies situated on U.S. soil. States and cities are spending hundreds of millions of dollars annually in programs and incentives to attract this direct investment to their jurisdictions to create jobs, diversify their economic base, and enhance tax revenues.

Because of the lack of economic diversity in several states and many municipalities, however, U.S. direct investment abroad has received mixed reviews from noncentral government leaders. Some consider that such investment is indispensable if U.S. firms are to remain globally competitive and believe that in the long run more jobs will be created locally by enterprises that have well-established international networks. Approximately 2,000 American companies have established 21,000 foreign subsidiaries in more than 120 countries. These subsidiaries annually manufacture and sell abroad products worth about $700 billion (see *New York Times*, 1991b, p. C2; *Wall Street Journal*, 1991, p. A16).

Conversely, other noncentral government leaders are convinced that U.S. corporations are abandoning facilities in the United States for countries with cheap labor and minimal environmental and worker protection standards. In a vote at the June 1991 annual meeting of the U.S. Conference of Mayors in San Diego, the delegates refused to approve a resolution supporting the proposed NAFTA (*U.S. Mayor*, 1991). At its 1993 meeting in New York City, the Urban Economic Policy Committee of the U.S. Conference of Mayors finally approved a NAFTA resolution, but it was watered down with so many restrictive amendments that it was virtually meaningless.[9] In general, many of these delegates fear that companies would eliminate high-paying jobs in urban areas and transfer the work to Mexico, where wages and fringe benefits in the manufacturing sector are only one seventh of U.S. levels.[10]

Globalization has also precipitated a major restructuring in many industries and a growing recognition in the United States of the implications of the internationalization of production. With this in mind, Senator Bill Bradley of New Jersey has pinpointed four transformations that are having a powerful impact on the U.S. economy: (a) the end of the Cold War, which has cost jobs in defense-related industries; (b) international competition that has opened markets for U.S. exports but that has also forced U.S. workers to compete with those abroad who generally receive lower wages; (c) an increased demand in the workplace for knowledge and technology, a phenomenon that has left some low-skilled, poorly educated Americans behind; and (d) the drive to lower the federal budget deficit, which has meant that the government is spending less on infrastructure projects that would normally produce more jobs (*New York Times*, 1993c).

Although the United States created approximately 20 million net new jobs during the 92-month economic expansion that began in November 1982 and ended in July 1990, almost 3 million jobs have been lost in the manufacturing sector since 1979, the peak year of manufacturing employment. States and cities have all had to cope with the costs and benefits of restructuring. Detroit has been badly hurt by the downturn in the domestic automotive industry, whereas Marysville (Ohio), Georgetown (Kentucky), Smyrna (Tennessee), Spartanburg (South Carolina), and Vance (Alabama) have already created or will create thousands of new jobs as a result of Honda, Toyota, Nissan, BMW, and Mercedes-Benz building huge assembly plants in their localities. With the marked decline in international petroleum prices in the early 1980s, Houston lost 225,000 jobs in a 5-year span before moving into recovery. A quarter of a century ago, New York City provided headquarters for 156 of the 500 largest industrial corporations in the United States. Today, slightly more than 50 maintain their headquarters in America's largest city ("The 500 by state," 1992, p. 290).[11] Between April 1989 and the beginning of 1993, recession and corporate restructuring were responsible for the New York City metropolitan area experiencing a net job loss exceeding 350,000, more than 10% of total employment in the region (Kagann, 1992, p. A14).

Immigration, another major intermestic issue, is also in the process of transforming several U.S. and Canadian cities. Approximately 1 million immigrants and refugees are now entering the United States legally each year, and 250,000 are entering Canada, with the vast majority in both nations settling in the 42 metropolitan areas with populations exceeding 1 million (see *Toronto Globe and Mail,* 1993, p. A2; King, 1990, p. 28). California had a net increase of 3.2 million new entrants into the state in the 1980s. Of these, 2.3 million came from other countries and located predominantly in the Los Angeles, San Francisco, and San Diego metropolitan areas (*New York Times,* 1991a, p. A14). During the same decade, according to the U.S. Immigration and Naturalization Service, New York City attracted 854,000 immigrants, mostly from developing countries. One third of all New York City residents are now foreign born, up from one fourth just a decade ago. Because of this immigration, New York City experienced a modest 3.5% population increase during the 1980s. Los Angeles added more people than any other U.S. city during the 1980s. Many of these new residents were immigrants. In both New York City and Los Angeles, the so-called minority groups now constitute a majority. This is also the case in 49 other U.S. cities with populations exceeding 100,000 (*New York Times,* 1992a, Pt. IV, p. 7; *USA Today,* 1992, p. 8A).[12]

The U.S. Immigration Act of 1990 also provides economic incentives for states and municipalities to attract a special category of immigrants. This act establishes up to 10,000 annual investor-immigrant visas for foreign residents and their families willing to invest large amounts of money in U.S. businesses and create jobs for Americans. Seven thousand of these visas require an investment of at least $1 million, and 3,000 an investment of at least $500,000 in high-poverty or rural areas. Although relatively few foreign investors have thus far chosen to take advantage of this new law, municipal governments are interested in attracting the lion's share of these immigrants. Many already sponsor state-authorized enterprise zones or soon will sponsor federally authorized empowerment zones that will qualify as investments in high-poverty areas.[13]

Several additional factors other than complex economic interdependence and the intermestic dimension of trade, investment, tourism, immigration, and financial linkages help explain the growing global involvement of noncentral governments. These include the uneven distribution of economic gains in America's vast federal system, electoral considerations, significant growth in noncentral governments plus an improved capacity to interact with international actors, and a willingness on the part of the national government to permit noncentral governments to strengthen linkages abroad.

On the other hand, this phenomenon is largely attributable to the traditional role of state and local governments linked to protecting their revenue base and to safeguarding the interests of the people whom they represent. The collective fiscal health of these noncentral governments in the United States is worrisome, with more than one half of the major cities struggling with significant budgetary problems during the early 1990s.[14] In addition, although the federal government provided a record $204 billion in transfer payments to state and local governments during the 1993 fiscal year, this represented almost a 20% decline in constant dollars earmarked for the cities since 1980, with per capita federal aid to municipalities falling from $63 per capita in 1980 to $29 in 1993, more than a 50% decline (*Nation's Cities Weekly,* 1993b, pp. 6-7). The problem is compounded further by the federal government's propensity in recent years to mandate new responsibilities to state and local governments without transferring adequate funds to pay for these predominantly social welfare programs, a chief cause of fiscal instability at the municipal level.[15] Moreover, the money that is made available by Washington often has many strings attached, and only a small portion can be used for economic development purposes. Indeed, 60 cents of each dollar of federal grant money are now earmarked for only two welfare programs—Aid to Families With Dependent Children and Medicaid (a health insurance program for low-income individuals and families)—versus only 40 cents for these programs in 1980 (*Nation's Cities Weekly,* 1993c, p. 10; "Why State Budgets Are a Mess," 1991, p. 21).

Despite the myriad social problems faced at the local level, the prospects for a big increase in federal funding to municipal governments are dim. Although annual deficits have declined recently, the cumulative U.S.

federal debt now surpasses $4.6 trillion, with annual interest payments on this debt exceeding $217 billion in 1993—more than that year's combined personal savings of the American people ($189.9 billion; "National Income and Product Accounts," 1994, p. 18). Interest payments are one of the fastest growing budgetary allocations of the federal government. Money spent on this debt liability is about equal to the total transfers to state and local governments. Moreover, two thirds of the budgetary allocations of a typical U.S. state is now earmarked for only three programs: welfare, Medicaid, and corrections. In Canada, both the cumulative federal government debt and annual debt payments are proportionally larger than in the United States, and a relative diminution in transfer payments from Ottawa to the provincial governments has emerged as a divisive issue in Canada's federal system.

Local Cleavages

Municipalities are further hamstrung by the limitations placed by their state governments on their access to revenue sources. With rising costs associated with explosive increases in health-care costs, homelessness, AIDS, crime, infrastructure deterioration, and suburban flight, municipal governments are facing major fiscal challenges. Health care now absorbs 14% of America's GDP, far more than any other industrialized country. Despite this huge expenditure, 37.4 million Americans were without health insurance at the end of 1992 and tens of millions of additional people had inadequate levels of insurance.[16] The U.S. infant mortality rate also ranks near the bottom of the industrialized countries, although the overall U.S. rate fell to a record low of 8.5 per 1,000 births in 1992—but still remained twice as high for Blacks (*USA Today,* 1993b, p. D1).[17] Many urban centers have a much higher rate than the national average, with New York City's a third higher and Central Harlem's almost three times higher (*New York Times,* 1990, Sec. I, p. 11). Although estimates vary dramatically, homelessness has apparently increased significantly. Among these homeless people are tens of thousands of persons who are in desperate need of mental health care but who now wander aimlessly on urban streets. AIDS has also afflicted hundreds of thousands of young people, a disease that had not even been identified less than two decades ago.

In terms of crime, the U.S. inmate population doubled to more than 1 million during the 1980s. The United States has the highest incarceration rate among the Organization for Economic Cooperation and Development (OECD) nations. The average cost of incarcerating a juvenile now approaches $30,000 per year, appreciably more than the cost of educating a young man or woman at many prestigious universities.[18] A report released by the U.S. Senate Judiciary Committee described the United States as perhaps the most violent and self-destructive country in the world. Violent crime increased by 516% between 1960 and 1990, whereas the U.S. population as a whole increased by only 41% (*Washington Post National Weekly Edition,* 1991a, p. 25). The U.S. murder rate is 15 times higher than that of Japan and Great Britain. More than one half of federal prisoners are now serving time on drug-related charges.[19] Much of this crime is an inner-city problem, with the overall rate of rape, robbery, and assault among city dwellers almost twice as high than among rural residents and more than 50% higher than among suburban residents.[20] Minority groups have been especially decimated by lawlessness, with Blacks being six times more likely to be murdered than Whites.

Dropping out of school also contributes to higher poverty and social unrest, at a time when 37 million Americans are already impoverished, including one in four children under the age of 6.[21] Municipalities such as New York City are experiencing an overall 33% school dropout rate, with the rate among minority groups in certain cities approaching 50% (*Washington Post National Weekly Edition,* 1991b, p. 10). During the 1980s, the poverty rates of the 39 largest metropolitan areas averaged 18% in the inner-city areas versus 7.3% in their suburbs. The 1990 census also indicated that the 31 largest concentrations of poverty in the United States were in inner cities (Labich, 1994, pp. 88-89). The percentage of the poor who lived in female-headed households shot up from 29% in 1959 to more than 53% in 1990, and the number of people living in the ghettos (areas in which at least 40% of residents qualify as poor) rose from 3.7 million in 1970 to 5.6 million in 1980 and then to 10.4 million in 1990 (*Nation's Cities Weekly,* 1993a, p. 6; *New York Times,* 1993d, Sec. IV, p. 5).

Public infrastructure deterioration is frequently referred to as America's third great deficit (in addition to its government and current account deficits), with perhaps $50 billion in added spending needed each year for modernization and repairs. Once again, much of the deterioration that has occurred in the public infrastructure is found in the inner cities. Furthermore, the difficulty in paying for infrastructure and other improvements at the municipal level has been exacerbated by the flight of the middle class to the suburbs. Many of these people work in urban centers and take advantage of city services, but their contribution to the municipal tax base is minimal. The per capita income in America's largest central cities is three fifths of that in their suburbs, whereas city-suburban income levels were roughly equal just two decades ago (*Nation's Cities Weekly,* 1991a, p. 6; *Washington Post National Weekly Edition,* 1991a, p. 6). The development of economically self-contained "edge cities" has also added to the miseries of the urban centers (see Garreau, 1991).[22] Moreover, during the recession of the early 1990s, those metropolitan regions with the largest income differentials between inner city and suburbs tended to suffer the greatest economic downturns.[23]

How America copes with the problems of its urban centers will have a dramatic effect on U.S. competitiveness in the 21st century. Whereas two thirds of all Americans lived in rural areas in 1890, three of every four now live in metropolitan areas (a population exceeding 50,000), and, as mentioned earlier, more than one half are concentrated in 39 metropolitan areas having at least 1 million residents.[24] Fully 90% of America's population growth during the 1980s occurred in these areas with more than 1 million inhabitants.

North American Cities and Global Economic Competitiveness

Although the following observations are geared specifically to U.S. municipalities, many should also be applicable to cities in Canada and Mexico. Furthermore, these observations focus on economic development, but they represent little more than a partial solution to the myriad problems facing municipalities that were outlined in the previous section of this chapter.

The United States is the world's largest economy with a $6.3 trillion GDP in 1993, representing almost one quarter of total global production. Moreover, NAFTA is a free trade area running from the Yukon to the Yucatán and is far larger in land mass than either Russia or China, with a combined population of 375 million people, an aggregate annual GDP of $7 trillion, and three-way merchandise trade approaching $300 billion in 1993.

As documented earlier, just as economic linkages are being solidified continentally, so are such linkages proliferating globally. How does the United States stack up in this increasingly interdependent global economy? It remains the world's largest trading nation, with $1.05 trillion in merchandise imports and exports in 1993, up from $466 billion in 1980. Its exports and imports of services also attained a record level of $318 billion in 1993, more than triple the $89 billion tallied in 1980 (Bureau of Economic Analysis, 1993, pp. 60, 63; 1994a, p. 63). In 1993, the amount of U.S.

goods and services sold abroad was almost 40 times greater than the amount sold in 1955.[25] The United States has long been the top-ranked international direct investor with an estimated $535 billion in cumulate IDI at the end of 1993, but it is also the largest host nation for IDI, with $441 billion in 1993, up more than 500% since 1980 (see Bureau of Economic Analysis, 1994b, p. 66; *Statistical Abstract of the United States,* 1993, p. 799).[26] In addition, international tourism has been the largest earner of foreign exchange for the United States during the past several years. In 1980, 22.3 million foreign residents visited the United States and spent $10.6 billion. In 1994, an estimated 48 million foreigners were expected to visit the United States and to spend $80 billion, providing America with a substantial international travel and tourism surplus (see *USA Today,* 1994, p. D1).[27] Exports and imports of goods and services, IDI, and tourism now account for approximately 15 million civilian jobs in the United States. Collectively, these international activities have been responsible for creating many new employment opportunities even during the slow-growth period of the early 1990s.[28]

Cities in the United States are on the front lines in this battle to be competitive at home and abroad. As Kresl (1992) has pointed out, in "this new highly internationalized environment, cities must compete aggressively against each other. They compete for major conventions, for location of production facilities, for headquarters' activities, for location of international organizations, for transportation connections or hubs, and for bridge or point-of-access city status" (p. 196). Moreover, because business competitiveness continentally and internationally is affecting the economic well-being of municipalities, one must give more credence to the notion of "thinking globally and acting locally." This maxim also indicates that an attitudinal change may be needed at the local level among government representatives, business executives, labor leaders, educators, and other important actors. Instead of being insular and dreading foreign competition, urban leaders must face it squarely and learn how to compete more effectively in an era in which goods can move through international airports on any two continents in a single day and financial and other business-related information can be transmitted around the world instantaneously. Indeed, much of the growth in the U.S. economy during the past 5 years is attributable to increased exports. Manufactured goods also accounted for more than 80% of U.S. merchandise exports in 1992, and workers who produce these goods usually enjoy above average wages. One in six manufacturing jobs is linked to exports, but only one third of manufacturers, many of them urban based, currently engage in export activity. The local business community is already battle tested in competing against domestic companies in the world's largest marketplace. Municipal leaders must now be prepared to assist these businesses in their efforts to compete beyond the national borders of the United States.

A business that is not nationally, continentally, or globally competitive today will most likely be the dinosaur of the 21st century because three quarters of everything made in the United States is already subject to competition at home and abroad from foreign producers. Protectionism emanating from Washington, D.C., a state capital, or a city hall will only prolong the agony of noncompetitive businesses and will also harm the interests of competitive enterprises that might face retaliatory protectionism in other countries.

The issue of municipal economic competitiveness in an interconnected global society has received considerable attention by researchers. After reviewing this literature and conducting interviews with city officials and specialists in this field throughout North America, Asia, Europe, and South America, it has become clear to me that the following 10 conclusions form the basis of a coherent set of factors that city economic planners should seriously consider when formulating their strategic responses to the challenges posed by economic internationalization.

1. The key role of small and midsized businesses. The economic well-being of municipalities will depend increasingly on the international competitiveness of small and midsized businesses. Fortune 500 corporations rarely need the assistance of any level of government in the export or investment domains. Moreover, these large corporations have shed 4 million jobs in the United States since 1980, whereas small and midsized companies have created more than 20 million net new jobs during the same period.

2. Emphasis on creating an attractive fiscal and regulatory climate for the existing business community. One may dream of attracting a Toyota assembly plant, as is the case with Georgetown (Kentucky), or a Nissan plant, as occurred in Smyrna (Tennessee), or more than 80 foreign companies, as has happened in the city and county of Spartanburg (South Carolina). The first step, however, must be to help the existing business community to prosper through enlightened regulations, competitive taxes, a modern infrastructure, respected educational and research facilities, and a high quality of life. These enviable conditions in turn will attract the attention of companies around the United States and indeed around the world. Thus, an indispensable priority for municipal governments is to work closely with the representatives of existing businesses and to spur on local entrepreneurial innovation.

3. A candid assessment of comparative advantages and disadvantages at the local level. During the past few years, the economies of the inland U.S. states have grown more rapidly than the economies of the coastal states, putting to rest the myth that a city must be in the Sunbelt and preferably on the coast with large chunks of beachfront property to expand its economic base.

Nevertheless, one must be dispassionate in listing the comparative advantages and disadvantages of the municipality. The task will then be to maximize the advantages and mitigate the negative effects of the disadvantages. For example, Buffalo is often referred to as the snow capital of the United States, but it has gained an economic bonanza from being located next to the major manufacturing centers of Canada. The Canada-U.S. Free Trade Agreement (FTA) went into effect in 1989, and Buffalo has lured many Canadian companies because the cost of doing business is far less and the regulatory climate far more attractive in upper-state New York than in Toronto. Several inland Western cities may also be relatively isolated, but they have attracted companies from the West Coast and overseas because of competitive business costs—plus they provide ready access to recreational areas, an important quality-of-life consideration. In France, although the city of Lille is located in the blighted northern region once dominated by the coal industry, it has aggressively marketed itself as a hub city in the expanded EU and has enticed numerous enterprises from neighboring nations to set up facilities within the city's boundaries.

4. Intergovernmental cooperation. Municipal governments can never lead local businesses in a direction these businesses do not want to go, but they can help them understand the challenges and opportunities in a rapidly evolving continental and global economy. The federal government can also help through the 68 domestic offices of the U.S. and Foreign Commercial Service (a division of the International Trade Administration [ITA]), the 50 District Export Councils whose members are selected by the U.S. Secretary of Commerce, the U.S. Small Business Administration (SBA), the U.S. Export-Import Bank (Eximbank), and the Overseas Private Investment Corporation (OPIC). ITA personnel are available to participate in export seminars, and their organization maintains a database with export, licensing, joint venture, and related leads. The U.S. Foreign and Commercial Service will also facilitate U.S. business participation in international trade shows and trade missions. The Eximbank provides limited funding for companies wanting to enter export markets, and OPIC will insure overseas investments against losses due to revolution and repatriation restrictions, for example.

State governments expended almost $100 million in 1990 for international trade, investment, and tourism programs (excluding the cost of investment incentives and tax-exempt bonding). Most continue to maintain offices overseas and actively sponsor international trade and investment missions. Several states also provide funding for small and midsized businesses to export, primarily in the form of loan guarantees, and many also offer incentive packages to lure domestic and foreign enterprises to their areas of jurisdiction. Almost all state governments also sponsor "How to Export" seminars that address financial, transportation, legal, and other business-related issues at nominal or no cost to local companies. In Georgia, the state government also hosts an annual program that takes representatives of foreign embassies, consulates, and commercial groups to cities throughout the state to meet with municipal officials and observe firsthand the local business climate.

Core cities and their immediate suburbs must also work together and cooperate much more effectively in preparing the regional economy for international competition. Seattle, Atlanta, and Toronto have been among the leaders in the United States and Canada in promoting close regional economic cooperation (see Kresl, 1989, for a discussion of Toronto; see Peirce, 1993a, for a discussion of Seattle and Atlanta).

Municipal government representatives should network with federal, state, county, and neighboring municipal officials to ensure that the interests of their local business communities are adequately represented by these other levels of government. The National Governors' Association, the National Association of State Development Agencies, the U.S. Advisory Commission on Intergovernmental Relations, the National Association of Counties, the National League of Cities, and the U.S. Conference of Mayors are umbrella organizations that might also assist municipal representatives in developing strong vertical intergovernmental networks.

5. Public-private sector linkages. Cities that have succeeded in becoming more globally competitive have done so through extensive cooperation locally among government agencies and nongovernmental actors such as businesses, educational institutions, and private civic groups. For example, sister city arrangements and international exchanges among service groups such as Rotary and Kiwanis may pay economic dividends, heeding the maxim that "first we will become friends, and then we will do business." Such institutional linkages are best left in the hands of local citizens who are dedicated to the ideals of their organizations. Municipal officials should be involved, but they simply do not have the time nor resources to direct such exchanges on an ongoing basis.

Chambers of commerce or their functional equivalents must also be key players. The experiences of midsized municipalities in France and Germany indicate the vital importance of close cooperation between local government officials and designated representatives of the business community. In Stuttgart, for example, the chamber of commerce spearheads the city's and the region's successful apprenticeship and vocational education programs. In the state of Washington, the Greater Seattle Alliance acts as the nexus for both vertical and horizontal networking. Elected officials must never forget that the private sector will continue to create the great bulk of jobs at the municipal level in the United States. These private enterprises must play the instrumental role in helping the local economy to be globally competitive.

In addition, it is wasteful for the inner city and its immediate suburbs to compete against one another to secure opportunities in the international economy. Although involving major restructuring, consolidated governments on a regional basis may be the optimal solution to avoid the wasteful duplication of services and cutthroat economic competition.[29] At a minimum, however, government and business leaders must work together and not permit municipal political boundaries to frustrate their efforts to create "entrepreneurial city regions" (Peirce, 1993b, p. 2045).[30]

6. A strong educational base and worker training. Educational institutions must be integral players on the municipal team. At the primary and secondary levels, these institutions should be expected to provide a well-educated and skilled workforce, whereas postsecondary institutions could provide advanced worker training and research-and-development assistance. Quite frankly, the United States has fallen behind Western Europe and East Asia in the development of brainpower and technical skills, as illustrated by the vast majority of international tests administered during the past decade, as well as national surveys indicating that almost one half of American adults have a low literacy level and that two thirds of children in primary and secondary schools read below their grade levels.[31] A competitive business community in the United States will require skilled and well-educated employees. Increasingly, municipal governments and academic leaders must work together to meet this challenge and to provide properly trained workers on an industry-by-industry or even company-by-company basis.

7. The investment incentive trap. Some municipal governments will be tempted to join with their state governments to provide special incentives to entice companies to their region from other parts of the United States and the world. For example, the city of Georgetown and the state of Kentucky will provide Toyota with an estimated $305 million in incentives during a 20-year period for locating an assembly plant in that small city. The state of Alabama and some subnational jurisdictions within Alabama have pledged more than $300 million in incentives to Mercedes-Benz to build an assembly plant in Vance—a much higher cost per job than Kentucky's transaction with Toyota. More than 30 states made bids for the Mercedes-Benz facility. Illustrating how cities and states can get carried away with incentives, New York City and the state of New York have actually anted up hundreds of millions of dollars in incentives in a last-ditch effort to keep some of America's largest corporations from vacating the city (see Kagann, 1992; *Newsweek,* 1993; *New York Times,* 1992b, 1993a). These incentive packages often include outright grants, loans, loan guarantees, tax exemptions, infrastructure development, worker training, site preparation, and so forth. This is a precarious route to follow and inevitably disadvantages local businesses that must compete against these enterprises that are the recipients of government incentives.

Surveys of European firms that have established facilities in the United States indicate that incentive packages rank low on the list of reasons for selecting one site over another. Domestic firms share the same priorities and in a recent survey rank-ordered the factors that influence their corporate location decisions: (1) real estate costs, (2) labor force issues, (3) transportation, (4) real estate availability, (5) market access, (6) regulatory environment, (7) labor costs, (8) community image, (9) tax climate/costs, (10) utility services, (11) utility costs, (12) quality of life, (13) business services, (14) incentives, (15) education system and training infrastructure, (16) proximity to supplies and raw materials, and (17) university resources (see *Cincinnati Enquirer,* 1993).[32]

Spartanburg, South Carolina, is perhaps the preeminent example of a smaller municipality that is exceptionally successful in attracting scores of foreign firms. Beginning 30 years ago with a population base of 45,000, the city leaders joined with the chamber of commerce and other business representatives in a conscientious effort to diversify the local economy, which at the time was almost totally dependent on aging textile plants.

They decided to target more efficient textile companies in Europe and to look for companies in other sectors as well. Today, scores of foreign companies call Spartanburg home and tens of thousands of jobs have been created. BMW has recently decided to build a manufacturing facility worth $300 million in the county and will employ 1,000 workers. Another 20 companies that supply parts to BMW are expected to locate in the vicinity, adding 2,000 more jobs.

The general incentives given to BMW and other foreign-based companies by the municipal and state governments include the modernization of the infrastructure, which can also benefit the existing business community, and specialized worker training. If incentives are offered by municipal governments in conjunction with their state governments, they should generally be limited to infrastructure improvements, worker training, and research cooperation between the companies involved and the public university system. In addition, either enterprise or empowerment zones might provide special benefits for businesses willing to locate in targeted municipal areas. Nevertheless, the same benefits should also be offered to existing businesses in their efforts to expand their facilities and create more jobs.

8. International tourism as an economic development tool. Foreign residents are visiting the United States in record numbers and spending unprecedented amounts of money. Approximately 20% of all tourists visiting New York City come from other countries, and they spend about $2 billion annually.[33] The Las Vegas Convention and Visitors Authority now has representative offices in three countries and is endeavoring to make the city user-friendly for foreign visitors. Approximately 2.5 million foreigners traveled to Las Vegas in 1991, up 78% from 1987. The total economic impact on the local community was $1.2 billion in 1991, up 133% from 1987.[34] Attracting visitors from overseas will become more imperative for Las Vegas in the years ahead because it now has more hotel rooms to fill than any city in the United States, as well as 9 of America's 10 largest hotels.[35]

A tourist in Las Vegas often desires to visit the Los Angeles area, the Grand Canyon, and Zion and Bryce Canyon National Parks, an itinerary that involves four states and several municipalities. Similar multistate and multicity itineraries are common for visitors who first travel to the Pacific or Atlantic coasts or to parts of inland America. Many overseas tourists are also repeat visitors who are looking for less traditional itineraries. At current exchange rates and airfares, foreign tourism in the United States will remain at or near record levels. Consequently, municipal governments should use intergovernmental and private sector networking to attract these visitors and to make their cities user-friendly for those who speak different languages and who come from a potpourri of cultural backgrounds.

9. The internationalization process at the municipal level. Various checklists and guidelines are available to assist urban leaders to identify the international linkages that have already been established by the municipal government and the people and organizations it represents.[36] In particular, the leaders should recognize that social and cultural diversity at the local level can contribute substantially to this internationalization process.

10. The institutionalization of the internationalization process. A city's effort to adapt to a global economy is difficult and requires a great deal of work, cooperation, and patience. Some cities have made great progress under the leadership of a certain mayor or civic leader only to take a major step backward once these personalities leave the scene. To be successful, intergovernmental and private sector linkages must be institutionalized as much as possible. Leaders must recognize that no single municipal government, whether big or small, can accomplish this task alone.

Without doubt, the interdependence of the global, national, and local economies will continue to develop for the foreseeable future, so the quest to internationalize U.S. cities should remain a major priority. Nonetheless, some local economies do better than others, and municipal budgetary constraints and pressures may at times tempt new leaders to abandon the internationalization process. Organizations and networks that are strongly committed to the internationalization process and that remain in place even when individual leaders may change can help avoid this pitfall.

In conclusion, North American cities are increasingly affected by international economic conditions in the post-Cold War era. Public and private leaders at the level closest to the people should be expected to work together to enhance the economic well-being of their constituents by fostering globally competitive enterprises. "Think globally and act locally" will undoubtedly be a municipal rallying cry in 21st-century North America.

NOTES

1. The United States has approximately 35,000 cities and towns, Canada 2,000, and Mexico 2,500. See Brunn, Yeates, and Ziegler (1993, p. 78); Instituto Nacional de Estadística, Geografía e Informática (1992, p. 6); Law (1992, pp. 19-1 to 19-44). In general, these statistics do not include villages, hamlets, and other sparsely populated units. The five territories

include the Northwest Territories and the Yukon in Canada, plus Puerto Rico, Guam, and the U.S. Virgin Islands in the United States.

2. Urban areas in this case are defined as towns or cities with at least 5,000 inhabitants.

3. With the expected entry of Austria, Finland, Norway, and Sweden into the EU in 1995, the population base expanded to 370 million people and the combined GDP to $7.7 trillion.

4. These figures were derived from a United Nations Council for Trade and Development (UNCTAD) study completed in 1993. See *Time* (1993).

5. British Columbia's Trade Development Corporation closed its offices in Singapore, Hong Kong, Taiwan, and Germany, stating that they were too expensive and too much of a plum for former politicians and favored bureaucrats. It will continue to operate offices in Great Britain, South Korea, and Japan.

6. In 1989, governors from 41 of the 53 U.S. states and territories made 82 trips to 35 countries. Forty-eight trips were primarily to encourage exports and 32 to increase investment. Japan was the leading country visited, with 19 trips, followed by Belgium, with 13 trips.

7. Zelinsky (1990) discusses in particular a business alliance between Louisville (Kentucky) and Montpellier (France) and business linkages between Baltimore (Maryland) and Xiamen (China) and between Portland (Oregon) and Sapporo (Japan).

8. Sources also include interviews with officials of the Las Vegas Convention and Visitors Authority and the San Diego Convention and Visitors Bureau during the summers of 1993 and 1994.

9. NAFTA was approved by the committee subject to the following conditions: (a) the inclusion of supplemental agreements that will create high-wage and high-skill jobs, improve the environment, and maintain strong health and safety standards; (b) the addition of a comprehensive worker assistance package to help those who may be negatively affected by increased competition, base closings, and other dislocations; (c) a commitment to raise environmental and labor standards in all three nations; (d) the elimination of foreign tax credits and deferrals; and (e) the denial of trade benefits to companies that violate policies designed to protect U.S. jobs and the environment. See *U.S. Mayor* (1993a).

10. Instead of focusing on wages alone, subnational leaders should concentrate on unit labor costs, which take into account wages plus worker training, worker turnover, worker absenteeism, the cost of capital, the sophistication of the infrastructure, and so forth. When this is done, most U.S. industries are more competitive or at least as competitive as their counterparts in Mexico.

11. Fifty-six of the top 500 industrial corporations maintained headquarters in New York City at the end of 1991.

12. In 1992, 641,000 students were enrolled in Los Angeles's public school system. More than 70% of the parents of these students were born overseas, and nearly one half of the students did not speak English as a first language.

13. In the federal budget approved by the U.S. Congress in August 1993, the Clinton administration was successful in its efforts to fund both urban and rural-based empowerment zones that will provide $2.5 billion in tax breaks and $1 billion in spending grants to investors willing to locate businesses within the zones. See *U.S. Mayor* (1993b, p. 6).

14. In a National League of Cities survey of 688 municipalities that was released in July 1993, 53% stated that current expenditures were expected to exceed revenues, with nearly 20% stressing that the gap would be greater than 5% of total revenues in 1993. When asked to provide the three most significant factors creating difficulties for their cities' fiscal conditions, the respondents most frequently mentioned city employee health benefits, infrastructure needs, and unfunded federal and state mandates. See *Nation's Cities Weekly* (1993b, pp. 6-7).

15. These federal mandates without adequate funding are for water treatment, public health, waste disposal, transportation, housing, education, and welfare. Mayor Richard Daley estimates these unfunded or underfunded mandates cost Chicago $160 million per year. See *U.S. Mayor* (1993b, p. 14).

16. According to a U.S. Census Bureau study, the number of persons without health insurance increased by 2 million between 1991 and 1992. See *New York Times* (1993b, p. A10).

17. These statistics were released by the U.S. Department of Health and Human Services.

18. In comparison, annual room, board, and tuition at Harvard University in 1991 was $18,000. See Raspberry (1991).

19. According to the U.S. Office of National Drug Control Policy, Americans spent an estimated $40 billion on illegal drugs in 1990, compared with $50 billion in 1989 and almost $52 billion in 1988.

20. These statistics were included in a U.S. Department of Justice study for the period 1987 to 1989. See the *Deseret News* (1992, p. A5).

21. According to U.S. Census Bureau statistics released in October 1993, 14.7% of the U.S. population lived in poverty in 1992, including 1 in 5 children and 1 in 4 preschool children. Among Blacks, the poverty rate was 33.3%; among Hispanics, 29.3%. See *New York Times* (1993b, p. A10).

22. Garreau (1991) suggests that more than 200 such edge cities exist, almost all larger than the cores they surround. For example, 92% of the people in the New York City metropolitan area live outside Manhattan (p. 6).

23. This conclusion is based on a study by the Federal Reserve Bank of Philadelphia. In contrast, the better the inner city fares economically, the better the suburbs will also fare. See Peirce (1993a).

24. In contrast, only 14 metropolitan areas had populations exceeding 1 million in 1950, and they represented only 30% of the U.S. population. See Peirce (1993a, p. 4).

25. The United States exported $17.6 billion in goods and services in 1955, whereas similar sales reached almost $650 billion in 1993. These dollar figures have not been adjusted for inflation. See *Business America* (1993a, p. 6).

26. These IDI statistics are calculated using the historical cost basis.

27. These estimates were made by the U.S. Travel and Tourism Administration.

28. According to U.S. Department of Commerce calculations, foreign-owned companies in the United States provide about 5 million jobs. Merchandise exports from the United States in 1990 were responsible for more than 7 million jobs, and one in six jobs in manufacturing is supported directly or indirectly by exports. Each $1 billion in merchandise exports will create on average 19,100 jobs in the United States.

Although the job creation figures for exports of services and tourism by foreign visitors are more difficult to calculate, the $180 billion in U.S. exports in services and the $70 billion in spending by overseas visitors to the United States, both of which occurred in 1992, should conservatively provide at least 3 million additional employment opportunities.

29. A persuasive case for the consolidation of contiguous urban and suburban governments is made in Rusk (1993).

30. Peirce (1993b) emphasizes that functional relationships are more important than official borders.

31. These results were reported by the National Center for Education Statistics and the National Assessment of Educational Progress. See the *Wall Street Journal* (1993, p. A11); *USA Today* (1993a, p. A1); and *USA Today* (1993b, p. D1).

32. This survey was administered to 1,200 corporate relocation executives by the Deloitte & Touche Realty Consulting Group. See the *Cincinnati Enquirer* (1993, p. G4).

33. In 1989, 25 million tourists visited New York City, with 1 in 5 being a foreign resident. Collectively, these tourists spent about $9 billion in the metropolitan area.

34. These statistics were derived from interviews with representatives of the Las Vegas Convention and Visitors Authority during the summer of 1993.

35. Las Vegas had 87,000 hotel rooms in 1994, compared with about 70,000 for New York City and 80,000 for the combined Los Angeles and Long Beach region. For a related article, see *USA Today* (1993c, p. 2B).

36. A set of guidelines is featured in Fry, Radebaugh, and Soldatos (1989) and Soldatos (1991a, 1991b). Furthermore, in Peirce (1993a), readers are provided with case studies of Phoenix, Atlanta, Seattle, Baltimore, Owensboro (Kentucky), Dallas, and St. Paul.

REFERENCES

Brunn, S. D., Yeates, M. H., & Ziegler, D. J. (1993). Cities of the United States and Canada. In S. D. Brunn & J. F. Williams (Eds.), *Cities of the world* (2nd ed., pp. 43-83). New York: HarperCollins.

Bureau of Economic Analysis. (1993, June). *Survey of current business.* Washington, DC: Author.

Bureau of Economic Analysis. (1994a, March). *Survey of current business.* Washington, DC: Author.

Bureau of Economic Analysis. (1994b, April). *Survey of current business.* Washington, DC: Author.

Business America. (1993a, May 31). p. 6.

Business America. (1993b, World Trade Week Edition). pp. 8, 22.

Camilleri, J. A., & Falk, J. (1992). *The end of sovereignty? The politics of a shrinking and fragmenting world.* Aldershot, UK: Edward Elgar.

Cincinnati Enquirer. (1993, September 19). p. G4.

Deseret News. (1992, June 8). p. A5.

Drucker, P. F. (1990). *The new realities.* New Delhi, India: Asian Books.

The 500 by state. (1992, April 20). *Fortune, 125,* 290-312.

Fry, E. H. (1989). The impact of federalism on the development of international economic relations: Lessons from the United States and Canada. *Australian Outlook: The Journal of International Affairs, 43,* 19-25.

Fry, E. H., Radebaugh, L. H., & Soldatos, P. (Eds.). (1989). *The new international cities era: The global activities of North American municipal governments.* Provo, UT: Brigham Young University, David M. Kennedy Center for International Studies.

Garreau, J. (1991). *Edge city: Life on the new frontier.* New York: Doubleday.

Governor's Weekly Bulletin. (1990, March 30). p. 1.

Instituto Nacional de Estadística, Geografía y Informática. (1992). *Anuario estadístico de los Estados Unidos Mexicanos, 1991* [Statistical yearbook of the United Mexican States, 1991]. Aguascalientes, Mexico: Author.

International Herald Tribune. (1993a, July 6). p. 4.

International Herald Tribune. (1993b, August 14). p. 9.

Kagann, S. (1992, October 6). New York's incentives, the wrong incentives. *Wall Street Journal,* p. A14.

King, A. D. (1990). *Global cities: Post-imperialism and the internationalization of London.* London: Routledge.

Knight, R. V. (1989). The emergent global society. In R. V. Knight & G. Gappert (Eds.), *Cities in a global society* (Urban Affairs Annual Review, Vol. 35, pp. 24-44). Newbury Park, CA: Sage.

Kresl, P. K. (1989). Variations on a theme: Internationalization of "second cities." In E. H. Fry, L. H. Radebaugh, & P. Soldatos (Eds.), *The new international cities era: The global activities of North American municipal governments* (pp. 185-198). Provo, UT: Brigham Young University, David M. Kennedy Center for International Studies.

Kresl, P. K. (1992). *The urban economy and regional trade liberalization.* New York: Praeger.

Labich, K. (1994, June 27). The geography of an emerging America. *Fortune, 129,* 88-89.

Law, B. (Ed.). (1992). *1993 corpus almanac and Canadian sourcebook.* Don Mills, ON: Southam Information and Technology Group.

National income and product accounts. (1994, April). In *Survey of current business* (p. 18). Washington, DC: Bureau of Economic Analysis.

Nation's Cities Weekly. (1991a, June 17). p. 6.

Nation's Cities Weekly. (1991b, July 1). p. 6.

Nation's Cities Weekly. (1993a, June 14). p. 6.

Nation's Cities Weekly. (1993b, July 12). pp. 6-7.

Nation's Cities Weekly. (1993c, September 20). p. 10.

Newsweek. (1993, May 24). p. 43.

New York Times. (1990, June 30). Sec. I, p. 11.

New York Times. (1991a, February 21). p. A14.

New York Times. (1991b, February 26). p. C2.

New York Times. (1992a, November 29). Sec. IV, p. 7.

New York Times. (1992b, November 30). p. C6.

New York Times. (1993a, October 4). p. A10.

New York Times. (1993b, October 5). p. A10.

New York Times. (1993c, October 7). p. A12.

New York Times. (1993d, October 10). Sec. IV, p. 5.

Peirce, N. R. (1993a). *Citistates: How urban America can prosper in a competitive world.* Washington, DC: Seven Locks Press.

Peirce, N. R. (1993b, August 14). There's a new way to define regions. *National Journal,* p. 2045.

Raspberry, W. (1991, May 13). Prison costs more than Harvard. *Washington Post,* p. A11.

Rusk, D. (1993, September 8). Suburban renewal. *New York Times,* p. A19.

Soldatos, P. (1991a). *Les nouvelles villes internationales: profil et planification stratîgique* [The new international cities: Profiles and strategic planning]. Aix-en-Provence, France: SERDECO.

Soldatos, P. (1991b). Strategic cities alliances: An added value to the innovative making of an international city. *Ekistics, 58*(350-351), 346-350.

Statistical abstract of the United States. (1993). Washington, DC: Government Printing Office.

Time (International Edition). (1993, August 2). p. 11.

Toronto Globe and Mail. (1993, September 16). pp. A2, B8.

USA Today. (1992, December 4). p. 8A.

USA Today. (1993a, September 9). p. A1.

USA Today. (1993b, September 16). p. D1.

USA Today. (1993c, October 7). p. 2B.
USA Today. (1994, February 8). p. D1.
U.S. Immigration Act of 1990, 104 Stat. 4978.
U.S. Mayor. (1991, July 1). pp. 8, 11.
U.S. Mayor. (1993a, June 28). p. 11.
U.S. Mayor. (1993b, August 16). pp. 6, 14.
Wall Street Journal. (1991, March 27). p. A16.
Wall Street Journal. (1993, September 16). p. A11.
Washington Post National Weekly Edition. (1991a, April 8). pp. 6, 25.
Washington Post National Weekly Edition. (1991b, May 27). p. 10.
Why state budgets are a mess. (1991, June 3). *Fortune, 123,* 21-30.
Williams, J. F., & Brunn, S. D. (1993). World urban development. In S. D. Brunn & J. F. Williams (Eds.), *Cities of the world* (2nd ed., pp. 1-41). New York: HarperCollins.
Zelinsky, W. (1990). Sister city alliances. *American Demographics, 12,* 42-45.

2 The Determinants of Urban Competitiveness: A Survey

PETER KARL KRESL

The concept of competitiveness has been widely discussed during the past two decades. During the 1970s, many economists became concerned about the ability of mature industrial economies to retain their positions at the apex of the economic hierarchy of nations. Britain had suffered from balance of payments weakness since the end of World War II. After experiencing trade surpluses from 1893 to 1971, the United States had just begun an unbroken series of balance of trade deficits that continues until today. The mood in the industrial economies of Western Europe was swinging from Europtimism through Europessimism, Europhoria, and Eurosclerosis to Eurorealism.

The situation was not improved when we moved into the decade of the 1980s. The Reagan-Thatcher years provided an emotional boost to broad sectors of the Anglo-Saxon world and a substantial increase in incomes to people at the upper end of the income distribution but, alas, no real gains in aggregate per capita incomes or production. The excitement during the first half of the decade at having a currency that was increasing in purchasing power relative to those of competing economies turned to gloom on realization, during the second half, that a strong currency tended to produce a substantial trade deficit and a weakened industrial sector. Europe had become inured to stagnation, but in North America the prospect that children would have economic lives less satisfying than their parents was a bitter pill to swallow. Hence the plethora of studies aimed at (a) explaining this period of secular stagnation and (b) providing a strategy to enhance competitiveness.

At the same time, there was a second sea change occurring in the international economy. After several rounds of tariff reduction through the General Agreement on Tariffs and Trade (GATT), further progress at trade liberalization on the global level became exceedingly difficult. In the mid-1980s, both Europe and North America turned their attention to regional liberalization, for which the prospect of additional gains appeared to be more promising. Both the completion of the internal market process of the European Community (EC 1992) and the Canada-United States Free Trade Agreement, soon followed with the addition of Mexico by the North American Free Trade Agreement (NAFTA), brought progress in areas that had proved to be too difficult for the GATT process—services, agriculture, subsidies, and, most significant, dispute resolution mechanisms. As a consequence of this progress, nations found themselves constrained in their ability to manage the international economic relations to achieve domestic policy objectives, such as full employment and price stability. Free recourse to interventions such as subsidies, blatant manipulation of the exchange rate, and bureaucratic measures designed to serve as a barrier to imports were now more likely to result in retaliation or referral to an impartial dispute resolution panel than in a short-term boost in incomes or employment of a nation's citizenry.

As nations moved with slow but sure steps toward the sidelines, regions and, particularly in North America, cities became both more significant as economic actors and more aware of the greater burden on them to strategically plan for their economic futures in an environment in which they were increasingly exposed to challenges to existing economic activities and to opportunities for growth into new areas emanating from the international economy. Readers will be acutely aware of this and will, indeed, have read descriptions and explanations of this phenomenon in the literature (e.g., the papers from the first New International Cities Era [NICE] conference: Fry, Radebaugh, & Soldatos, 1989; see also Kresl, 1992; Peirce, 1993). They will also be aware that although there is a considerable literature dealing with the concept of competitiveness from the traditional standpoint of the national economy, little has been written on the notion of the competitiveness of the individual city, standing as a relatively free agent in a redefined economic space.

In this chapter, I first discuss the concept of competitiveness as it has been developed in international economics. Second, I will consider competitiveness as it could be applied to an urban economy in the context of a highly competitive environment that extends beyond the regional or national economy to that of the continent and beyond. The final section of this chapter will confront the task of making this concept of the international competitiveness of cities operational. With an operational conceptualization of a city's international competitiveness, one could (a) evaluate and rate the position of an individual urban economy and (b) suggest a strategy for the individual city that seeks to enhance its competitiveness (see Kresl & Singh, 1994).

Competitiveness: From the National to the Subnational Economy

Because the nation is not the focus of this chapter, I will discuss only four of the contributions to the competitiveness literature; furthermore, there tends to be agreement among economists on most core notions. For many economists, particularly of the political left, the perceived loss of national competitiveness of some industrial economies that became apparent in the 1970s was an indication of a process of deindustrialization, whereby a previously viable, export-surplus generating manufacturing sector was transformed into a declining sector of the economy, and the trade surplus was turned to persistent deficit. One such economist was Singh (1977). Rather than using the term *competitiveness,* Singh stresses the importance of what he refers to as "an efficient manufacturing sector" (p. 128). Although many writers complacently note a "natural" evolution from manufacturing to services, similar to that from agriculture to manufacturing, Singh argues that services will not be able to pick up the burden borne by manufacturing for providing exports (the base from which growth of services would occur is too small) and employment (the loss of jobs from a declining manufacturing sector has not in practice been offset by increases in jobs in services, especially in jobs of equal or higher quality or desirability). His definition is this:

> An efficient manufacturing sector [is] one which *(currently as well as potentially) not only satisfies the demands of consumers at home, but is also able to sell enough of its products abroad to pay for the nation's import requirement . . . at socially acceptable levels of output, employment and the exchange rate.* (p. 128; italics in original)

Lacking such a manufacturing sector, the United Kingdom (UK) was facing a future of stagnation, high unemployment, and a falling standard of living. The causes of this situation were, according to Singh, "deficiencies on the supply side—a manufacturing sector which clearly responds rather ineffectively to changes in domestic and foreign demand" (p. 131). These dynamic deficiencies are shown by "slower technical progress, product innovation and ability to respond to demand changes as compared with competitors" (p. 133). Thus, for any economy—and for this chapter I must include any regional or urban economy—a healthy and dynamic manufacturing sector is essential for international competitiveness.

At the other end of the political spectrum, Scott and Lodge (1985) wrote the introduction to a massive study of competitiveness of the U.S. economy from the vantage point of a business school. In this introduction, they provide the definition of competitiveness that guided the work of over 150 contributors. For these writers, "national competitiveness refers to a country's ability to create, produce, distribute, and/or service product in international trade while earning rising returns on its resources" (p. 3), without, they add in a footnote, resorting to aid from import restraints or artificial subsidies. Scott and Lodge state specifically: "We have chosen the nation state—not the firm or industry—as the basic unit of analysis," reasoning that it is the full employment and standard of living of Americans, not the workers of foreign subsidiaries of U.S. firms, that are important (p. 3). Scott and Lodge agree with Singh (1977) that manufacturing remains the core for U.S. competitiveness because it provides higher wages and output per employee, 70% of exports (in 1980), and more than 50% of imports. Only in manufacturing are significant productivity gains to be found.

A third contribution to this debate that has relevance to this chapter is the MIT study *Made in America* by Dertouzos, Lester, and Solow (1989/1990). Although first published in 1989, the work behind this project was begun in 1986 when many analysts and journalists were focusing on the negative impact on U.S. competitiveness of recent macroeconomic policies that affected savings and investment and resulted in a 50% increase in the value of the dollar, which priced U.S. goods out of domestic and international markets. *Made in America* shifted the attention to microeconomics by demonstrating that U.S. trade performance was a composite of many firms that were failing and many that were succeeding very well: "The causes of this (productivity) problem go well beyond macroeconomic explanations of high capital costs and inadequate savings to the attitudinal and organizational weaknesses that pervade America's production system" (p. 166). The headings of the conceptual section of the book illustrate clearly their concerns: Outdated Strategies, Short Time Horizons, Technological Weaknesses in Development and Production, Neglect of Human Resources, Failures of Cooperation, and Government and Industry at Cross-Purposes.

Although not disputing the primary focus of the aforementioned writers on manufacturing, productivity, and innovation, Porter (1990) leads away from what must be taken as the consensus view of this issue, at least in the mid-1980s, in his seminal book, *The Competitive Advantage of Nations:*

> We must abandon the whole notion of a "competitive nation" as a term having much meaning for economic prosperity. The principal economic goal of a nation is to produce a high and rising standard of living for its citizens. . . . The only meaningful concept of competitiveness at the national level is national productivity. A rising standard of living depends on the capacity of a nation's firms to achieve high levels of productivity and to increase productivity over time. . . . A nation's firms must relentlessly improve productivity in existing industries by raising product quality, adding desirable features, improving product technology, or boosting production efficiency. (p. 6)

Porter then presents his by now well-known "diamond of determinants of competitive advantage" (pp. 158, 622) with four points highlighting (a) firm strategy, structure, and rivalry; (b) demand conditions; (c) factor conditions; and (d) related and supporting industries. Although the individual firm has a clear responsibility, as does the national government (by way of its responsibilities for regulation of competition, macroeconomic demand management, and infrastructure investments), his approach presents the opening for the emergence of cities as significant economic actors and determiners of competitive advantage. Porter makes this quite explicit:

> Internationally successful industries and industry clusters frequently concentrate in a city or region, and the bases for advantage are often intensely local. . . . While the national government has a role in upgrading industry, the role of state and local governments is potentially as great or greater. . . . The process of creating skills and the important influences on the rate of improvement and innovation are intensely local. (pp. 158, 622)

Thus, the city emerges as an important, in many cases decisive, agent in the determination of competitive advantage.

This potential for economic activism by cities is made all the more pressing by several changes in the international economy. First, trade liberalization both at the global (GATT) level and the regional level (EC 1992 and NAFTA) are making it more difficult for national governments to intervene economically than has historically been the case. Certain forms of policy behavior (imposition of a host of tariff and nontariff barriers) have been severely curtailed, and nations that do not conform find themselves brought before impartial dispute resolution panels. Second, major changes in the technology of production, distribution, communication, and transportation have dramatically altered economic space, forced a reconsideration of center and periphery, and made it rational for cities to form new linkages and functional networks with other cities that previously had been considered economically remote. Hence the growing need for research on the international competitiveness of cities.

Competitiveness and the Urban Economy

When the focus is clearly on the city, the notion of competitiveness must be altered to reflect the reality of the urban economy and its capacity for action. Many of the elements that are important for the nation, such as the balance of payments and trade, the exchange rate, aggregate demand management, and antitrust legislation, have no relevance for the city. Some cities, of course, are simply trying to provide any type of employment possible to workers who have been laid off by firms incapable of meeting competition from newer or more efficient producers. Other cities, however, are in the more enviable position of restructuring their economic activity to improve the lives of the citizenry. Although in many ways less pressing, this latter challenge is more relevant to the focus of this chapter because it forces cities to make choices about their economic futures and to plan to create a competitive advantage.

Although the primary subject of this chapter will be an examination of the determinants of urban economic competitiveness, it will be useful to begin with consideration of the indicators of competitiveness and the characteristics of competitiveness. The indicators suggest desired results that will be increasingly present as cities are relatively more competitive. The characteristics are features of an urban economy that are concomitants of competitiveness.

Indicators of Competitiveness

Competitiveness is not just a question of promoting more growth but rather a process of economic evolution that will generate specific results that are considered especially desirable. For a relatively competitive urban economy, more jobs may be less important than "better" jobs, production should be of certain types of goods, certain activities will have desired long-term impacts, and so forth. If the following are present, the urban economy is relatively competitive:

- The jobs created should be high-skill, high-income jobs.
- Production should evolve toward environmentally benign goods and services.
- Production should be concentrated in goods and services with desirable characteristics, such as a high income elasticity of demand.
- The rate of economic growth should be appropriate to achieve full employment without generating the negative aspects of overstressed markets.
- The city should specialize in activities that will enable it to gain control over its future, that is, to choose among alternative futures rather than passively accepting its lot.
- The city should be able to enhance its position in the urban hierarchy.

In other words, for the city that can strategically plan not only to achieve more than merely its own survival but also to enhance its competitiveness, this planning is clearly a question of qualitative as well as quantitative factors.

Determinants of Competitiveness

The more important concept is that of the elements that actually can be used to explain the concept of urban competitiveness itself, as well as to assess the relative degree of competitiveness of an individual city and to identify both its strengths and weaknesses. This latter aspect will be of crucial importance in suggesting a specific strategy for enhancing competitiveness of the individual city. In making this approach to city economic competitiveness analytically more operational, I suggest the following, rather intuitive, formulation:[1]

Urban Competitiveness = *f* (economic determinants + strategic determinants)

Where,

Economic determinants = factors of production + infrastructure + location + economic structure + urban amenities

Strategic determinants = governmental effectiveness + urban strategy + public-private sector cooperation + institutional flexibility

Although the distinction between economic and strategic determinants of competitiveness may seem somewhat forced at first glance, the real distinction lies in the possibility that the economic determinants may be represented by statistical data, whereas the strategic determinants will always be qualitative in nature.

The concept of a city's international competitiveness is quite different from the concept of an "international city." This latter concept has been articulated by Soldatos (1991) and is concerned primarily with evaluating the degree to which a city is in fact international in its connections with the rest of the world economy and with designing strategies to enhance that internationality. As will become clear, a city may dramatically increase its competitiveness, even its international competitiveness, without being or increasing the degree to which it is an international city.

Examination of the Literature on the Importance of Individual Determinants

A richer understanding of the nature and importance of the individual determinants of urban competitiveness can be gained through a review of what economists and other policy analysts have learned through their research. The objective of this review is to highlight aspects of these determinants over which the city can, through its strategic plan and other policy initiatives, exercise some influence: How can a city affect each of these determinants to enhance its international competitiveness? This literature search will be confined to publications focusing on urban and regional economics because these seem to be most germane to the subject of this chapter.

Economic Determinants

Factors of Production

It has long been recognized that high productivity and increasing productivity are keys to making an economy more competitive. This, of course, means one must pay attention to the quantity and quality of the labor force, the capital stock, and production and office sites. The literature on factors of production is enormous. This chapter will consider only a small number of directly relevant sources. Mullen and Williams (1990) get to the heart of the matter when they argue that it is the newness of the capital stock that is important because newer investments are endowed with the most recent technological advances. The productivity gains that result from the newer technology generate output growth and thus an increase in labor efficiency, in part because of the in-migration of skilled workers who are attracted to these better jobs. Thus, the right sort of capital investments has wide-ranging effects including those on the skill and educational levels of the labor force.

Beeson (1990) gives backhanded support for the importance of capital in her study of the decline of manufacturing in large cities. The primary cause of the slow relative growth rate of large standard metropolitan statistical areas (SMSAs) between 1959 and 1978 was found to be low relative growth rates in labor and productivity; the primary cause of the relative decline of manufacturing in the large SMSAs was a lower rate of capital accumulation. The work of Mullen and Williams (1990) confirms the linkage between the rate of capital accumulation and productivity.

In getting access to capital for new investments, it is not necessary for the accumulation process to be local. Florida and Smith (1992) give a clear mapping of where and how capital is collected in certain locales, specifically New York, San Francisco, Chicago, and Boston, and then invested in others—Silicon Valley and Route 128. Coinvestment is important in encouraging long-distance capital movements to a locale. Local governments have two ways to attract this capital. First, an adequate and growing high-tech infrastructure is needed by venture capital. Second, Florida and Smith stress "get[ting] back to the business of building integrated strategies to bolster the underlying economic and technological capacities of states, regions, and metropolitan areas" (p. 207).

In addition to the specific needs of investment in high-technology areas of production, an adequate financial sector will be more important for a region that seeks to increase imports than for one that emphasizes its exports. Goldberg, Helsley, and Levi (1989) note that exporters may face significant incentives to do their financing in the country in which they will be selling their goods, if for no other reason than to reduce their exposure to foreign exchange risks. They conclude that "trade shifts the location of financial activity away from regions dependent on exports toward regions dependent on imports" (p. 5).

The roles of the quality and the quantity of labor in enhancement of a city's international competitiveness must be seen as analogous to their roles in economic development. In a study of the economies of the north and the south in Italy, Pompili (1992) shows that "the distribution and dynamics of human capital explain both the generally backward position of the South and the differences in production structure and functional control within the rest of Italy" (p. 928). Within the north, he finds that

Milan's richness in technical, scientific, managerial, and entrepreneurial human capital stock makes it Italy's complete industrial center, followed closely by Turin. The other cities, such as Bologna, Genoa, Florence, Venice, and Rome, have strengths only in specific sectors. The historic question of causation is not answered, but Pompili clearly concludes that the existence of labor with certain skills is a powerful attraction to certain activities. He notes that the differentials among the cities seem to shrink over time, suggesting a "professional urban life-cycle" (p. 932), but this could be because of the efforts of cities to broaden their economic structure or because of the spreading of individual firms beyond their original base of operations.

The relationship between skilled labor and competitiveness is actually a rather rich one. Beeson (1992) stresses the role that large agglomerations have in developing this sort of a labor force through their impacts on the ability to learn. Her research leads her to conclude that (a) higher output per worker gives rise to more opportunity for learning by doing, (b) producing high-technology goods does the same, and (c) because "an individual's ability to learn depends on the average knowledge of those around him" (pp. 28-29), cities facilitate this process by placing workers in close proximity with a large number of other workers from whom they can learn. This suggests that a city can influence positively its competitiveness by introducing subsidies to improve the average level of education and training or through an industrial policy that subsidizes high-tech industries. Above all, it is vital that the city retain these workers once they are trained, and it must be remembered that this highly skilled labor force concentrates in large cities because they place a higher value than do other workers on the cultural, educational, and lifestyle amenities that only large cities can offer. Beeson offers hope to all cities seeking to attract a skilled labor force, not just the largest cities, by suggesting that once a minimum size of city has been attained, further growth may be marked primarily by increased disamenities and negative externalities, such as congestion and pollution.

This conclusion is supported by Calem and Carlino (1991), who indicate that agglomeration economies are present in SMSAs up to a population of about 2 million, but that for larger cities, constant returns are found. This leads them to argue that technical progress is introduced in large cities and "filters down the hierarchy of cities" (pp. 88-89). It also is consistent with the observation that manufacturing has tended to shift to smaller metropolitan places, at least during the 1970s.

Infrastructure

The infrastructure needs of a city trying to enhance its competitiveness are dependent on the specific role that city seeks to create for itself. Some cities may try to become point-of-access or bridge cities, taking advantage of their proximity to contiguous economy, whereas others may seek to develop a regional, national, or international vocation. They may wish to become export bases, to develop niches in certain industries, or to build on an existing capacity to become a center of research and development (see Kresl, 1992).

Some strategies will require investment in transportation infrastructure to move goods or people or in communications facilities to enable the actors in the local economy either to transmit decisions to other parts of the global economy or to receive commands from more significant centers of decision making. It may be necessary to promote development of one complex of business and financial service providers rather than another: Are the services to be provided to large firms or to small and midsized firms? A workforce with certain characteristics may require that a specific set of urban amenities (such as housing, educational, recreational, and cultural opportunities) be made available. Functional linkages and networks may have to be created for the city to be successful in establishing itself in a regional, national, or international context. Relationships between universities, firms, and research laboratories throughout the economic space may be essential to success. Thus, each city will have to determine the complex of investments and initiatives that will be most important for its own specific strategy.

Heenan (1977) reports that the most important elements for international businesspersons are the city's political stability and its supporting infrastructure. He speaks of "building indispensable cities" and refers to Walter Wriston and Henry Ford II as executives who have "supported the construction of whole new urban environments" in their cities, New York and Detroit. For the city in competition with other cities for economic activity, it is "performance, not price competition" that is important (pp. 84, 90, 91).

Moomaw and Williams (1991) argue that the most important determinant of variations in total factor productivity growth (which they link to rapid output growth) are investments by the state in education and in the transportation infrastructure. The city has a key role in determining the priorities for expenditures by all levels of government in these two areas, so their conclusion that "states can influence their manufacturing growth rates" (p. 32) can certainly be applied to cities as well. Heenan (1977) cautioned against using tax breaks and other incentives to attract business, and Moomaw and Williams (1991) find that tax increases will not discourage manufacturing growth if they are accompanied by productivity-enhancing expenditures by government. Taylor (1992) refers to results of Mill and Carlino (not available to me) showing that local taxes have little explanatory value regarding the long-run growth of total employment but that the relationship between employment growth and highway infrastructure is large and significant. The same result is given in O hUallacháin and Satterthwaite (1992), who add that "focused development incentives that emphasize infrastructure improvements designed for particular industries and firms" (p. 54) have positive effects on growth with particular stress on research parks and enterprise zones.

Manufacturing is not the only activity that is positively related to development of the appropriate infrastructure. Beyers (1992) studied metropolitan growth and found, as did Singh (1977), that the linkage between manufacturing and producer services was strong and that changes in business practices by goods-producing firms have been integral in the recent growth of the producer services sector. Beyond this intersectoral symbiotic relationship, there is much that the city or regional government can do. Beyers finds that in addition to maintaining the vitality of the region's or the city's economic base, a development policy must also "develop high-quality infrastructure in the telecommunications field" and "support the education and training system" (p. 145).

Location

Location theory has been at the heart of economic geography and urban and regional economics for as long as these disciplines have existed. The literature is enormous on such aspects as center and periphery, agglomeration effects, localization, and the influence of urbanization on economic decision making. The rapid development of new technologies of production, transportation, and communication are causing a profound reconsideration of the basic tenets of traditional theory. The advantage of the center over the periphery is now greatly reduced, and goods and services, information, and decisions can be transported longer distances with no loss in time, efficiency, or cost effectiveness. This has obvious implications for perceived notions of agglomeration, localization, and urbanization. Klaassen (1987) even states that this completely "upsets location theory" (p. 253).

In his study of regional policy in France, Hansen (1987) concluded:

> The industrial restructuring and diffusion of transportation and communications that have taken place since [1975] have the potential to transform information networks and industrial organization, which in turn could reverse the long-run tendency toward geographic concentration of production in favor of a more balanced development among the different regions. (p. 9)

He does caution, however, that this may not take place without public policy measures "that reinforce the endogenous potentials of peripheral regions" (p. 9). Krugman (1991) is of the opinion that the changes that relate to the 1992 process of the European Community "will actually favor peripheral manufacturing" (p. 98).

This places a considerable burden on the local government to implement the correct set of policy initiatives that will, in a sense, defeat distance. It is most unlikely that multinational firms can be relied on to make the decision to locate in any given urban economy, nor will the transfers of technology that are so vital to productivity gains and to output growth occur if left to chance (Harrison & Hart, 1990). Emphasis must be placed on development of institutions, such as science parks, that will facilitate access to "process innovations" (Bonomi, 1990, p. 5), and the city must ensure that the small firms have access to the infrastructure and supporting business and financial services vital to their development (Hansen, 1987).

It remains true that for many cities, location will continue to be the single most important factor in determination of their economic futures. Geographical features such as mountain ranges, rivers, ports, forests, and mineral deposits will give certain cities obvious advantages in any international competitive struggle. Some cities, such as Buffalo or Lyon, will have natural vocations as bridge cities between two markets or economic spaces. Others, for example, Toronto and Amsterdam, will serve as points of access or bases of operations for international or intercontinental penetration of a continent (Kresl, 1991). But most cities do not have a dominant locational characteristic and have considerable leeway regarding the most rational strategy for their economic future.

We are left with the realization that cities have a greater capacity than they have ever had to position themselves in new economic spaces, created by regional trade liberalization and technological change, in ways that challenge traditional notions of what is possible for a city with a given location. To be on the periphery in a spatial sense no longer means that a city must also be marginal in an economic sense.

Economic Structure

The economic structure of an urban economy is important for that city's competitiveness because it is one of the key determinants of the degree to which that economy will be able to (a) take advantage of the opportunities that present themselves, (b) gain access to foreign markets and technologies, (c) establish necessary linkages with others, and (d) provide growth in employment, production, and exports. The aspects of economic structure that seem to be most significant are the distribution of firms by size, the role played by foreign-owned firms, and the richness of the complex of firms providing business and financial support services. More than with the other determinants of competitiveness, the literature seems to be quite mixed as to the most desirable characteristics.

Regarding the first aspect, firm size, it has long been recognized that the substantial majority of U.S. exports are produced by the largest firms. This is certainly true of the major export goods—aircraft, organic chemicals, road vehicles, electrical machinery, and computing equipment. But economics deals with things at the margin, and today at the margin large companies are retrenching, restructuring, and resizing downward. This is true not only of firms that are losing markets and are in significant difficulty but also of healthy firms seeking to become leaner and more profitable.

For these reasons, increased attention has been given to small and midsized firms as sources of growth in employment, production, and exports. These firms have indeed been the origin of most of the job creation in recent years. It has also been argued that smaller firms are better able to position themselves to take advantage of niche markets, to exploit new technologies in production and, in general, to react quickly to rapidly changing competitive conditions. Hence, much publicity and effort has been devoted to encouraging the formation and viability of smaller firms and their capacity to export. Smaller firms lack the resources and experience with exporting to foreign markets that are common to most large firms. Nevertheless, some of the literature contends that cities and other levels of government can assist with export financing; first encounters with foreign exchange markets, foreign legal systems, and foreign languages; and contacts with the appropriate providers of the wide variety of business services that exporting requires.

Recently, however, analysts have begun to argue that this picture is not entirely true. Many small firms, in fact, perform work given to them by large firms, so that a healthy small-business sector may in some instances depend on the existence of a set of healthy large businesses. Some frustration is also expressed with the minimal impact in the short run that can reasonably be expected from the policies that have been suggested. The final counterargument, of course, is that many of the large firms of the future are today's smaller firms. Thus, although the bloom is off the rose, so to speak, when it comes to big results achieved quickly, city officials should continue to assist the viability of smaller firms.

The presence of foreign-owned firms is supported because many innovations in technology, new products, and advances in management are kept within the originating firm. If this is not the case, it is usually true that it is relatively costly to gain access to them in open market transactions. An economy with a high number of foreign-owned firms may then be one with privileged access to elements that are crucial to competitiveness. This issue was the subject of considerable research in Canada during the 1970s and 1980s in conjunction with the establishment in 1975 of the Foreign Investment Review Agency. At the time, more than 50% of Canada's manufacturing sector was foreign owned, with 85% of that U.S. ownership. Critics argued that foreign firms did not invest in research facilities in Canada, that they tended to import more and export less than did Canadian-owned firms, that they impeded the formation of new firms by Canadians, and that they raised funds locally rather than adding to the stock of capital in Canada. At the time, U.S. officials and businesspersons argued strongly against this view, but the same critique has been raised in recent years in their own backyard as foreign ownership of U.S. manufacturing has mounted.

The issues raised are not easily resolved, but research has confirmed that foreign firms in Canada tend to import more of their components and intermediate goods than do locally owned firms. Some research laboratories were established in Canada, but this was often under considerable pressure from government. It appears that many of the other anticipated negative impacts of foreign ownership were forestalled by governmental pressure and generous grants from all levels of government used to induce firms to "do the right thing."

A review of the available literature leads to the conclusion that in the competitive international economic environment of today, governments are correct to continue to view positively the establishment of, or inward investment by, foreign-owned firms. The literature indicates, however, that the use of tax breaks and heavy subsidization of that inward foreign investment to accomplish this objective is doubtful.

The last aspect of economic structure, the adequacy of the complex of providers of business and financial support services, is easier to resolve. All studies now demonstrate that this is an important item for a city that seeks to enhance or just to retain its international competitiveness. Previously, it was perceived that the larger firms were quite indifferent to the services provided in a city in which they were located because much of the work that service firms would provide was done internally. Now, however, large firms are paring their staffs and the functions they perform and are contracting these tasks out to other, usually smaller, firms. This accounts for much of the rapid growth in recent years in virtually all aspects of the service sector in virtually all locations. Smaller firms were, of course, always understood to be in need of independent providers of the services they themselves were not large and complex enough to provide internally. It is now generally recognized that a city that is lacking in competent and low-cost services sector firms is facing a considerable handicap in either attracting firms to its area or successfully promoting the formation of new firms by local entrepreneurs.

Urban Amenities

Urban amenities are those aspects of human society that can be realized only in cities—places in which there is sufficient aggregation of people to support a variety of relatively exotic services. The provision of "high culture" (museums, orchestras, opera and dance companies, galleries, and other performance and exhibition spaces) ranks high on this list, along with historic districts for residences, shopping and recreation, educational institutions, parks, ethnic neighborhoods, and so forth. Although these amenities are widely recognized for their contribution to a healthy tourism industry, they are perhaps more important for the effect they have on the local labor supply. Other variables considered by researchers to be amenities, such as temperate weather and closeness to mountains or an ocean, may be significant determinants of the ability of a city to advance itself as a prime location for desirable economic activity but are not conditions about which any individual city can do much. A city cannot decide to create them as part of a strategic plan; for this reason, these latter amenities will not be considered in this chapter.

The arts are properly taken to be central place functions because only in large cities can the necessary agglomeration economies be realized (Heilbrun, 1992). In addition to the need for a critical mass of creative people, both performers and artists and production workers (Kresl, 1989), the concentration of individuals and firms has substantial cost savings to all participants. Given their importance in the creation, expression, and maintenance of distinctive national cultures, the arts industries have to be concentrated in centers in which access to media, performance and exhibition space, and, perhaps most important of all, a large audience are present. Thus, it is difficult to separate the clustering of the arts from the clustering of the high-income working population one finds in decision-making and technology-based urban economies.

Klaassen (1987) stresses the importance of amenities in the residential preferences of high-skilled workers, such as those found in the Alpine region of Western Europe. Important for this area is its abundant mountain recreation, but Klaassen also stresses the region's seven international airports and its cultural events and facilities such as the Salzburg Music Festival, La Scala in Milan, and the new opera house in Lyon. He notes that this regional economy has been successful in attracting firms in industries for which access to resources is not important and in which sophisticated communication can replace the previously valued personal contact. In these industries, he argues, "the interests of the employee are becoming the decisive factor in the choice of location" (p. 253). Clark and Cosgrove (1991) support Klaassen's emphasis on the importance of amenities by showing that they are as important as labor market opportunities in explaining migration of workers. Workers will move considerable distances to gain access to certain climatic conditions (e.g., sun, warmth) as well as museums, major league sports, and a better airport, in addition to better wages. Although Clark and Cosgrove do not believe they can discriminate between the climatic and cultural conditions sufficiently to propose that municipal officials embark on initiatives to improve cultural facilities as a means of attracting migrants with desirable skills, in another study, Clark and Hunter (1992) do conclude that "policy makers may be able to at least partially offset the deterrent effect of a poor mix of climatic amenities through the development of cultural and recreational amenities"—as long as taxes are not raised (p. 363).

Van den Berg and Klaassen (1989) argue that with modern transportation and communications, firms have access to virtually all cities and towns for their plant locations and that this "calls forth greater economic competition among the towns." As a consequence, "cities need to develop a strategy of supplying better-quality housing, shops, cultural and leisure-time provisions" (p. 57). To sum up this influence, note Peirce's (1993) reference to Dallas consultant James Crupi when he stresses the increasing importance of the "symbiotic relationship between arts, culture, health care, crime, the environment and economic strength" (p. 308).

Strategic Determinants

Effective Governance

The need for effective governance is made clear by considering two extreme patterns of city governance. For example, Chicago is a city of more than 3 million inhabitants surrounded by about 50 towns and cities, only a couple of which have a population of more than 75,000. The sheer differences in size make any cooperation an "elephant and mouse" situation in which the smaller cities have little incentive to participate, given the potential benefits of being "free riders." Buffalo, because of the peculiarities of its location, has a well-thought-out plan for development, but the governance structure includes four bridge commissions, two airports, two counties, and the federal, state, and municipal governments—with no single individual or agency in command. Little evidence exists that in either situation a rational decision-making process is likely to emerge.

The counter is provided by the governance structure of the larger Canadian cities with metrowide authority for activities in which there are significant externalities, scale efficiencies, and common interests. The Montreal Urban Community, Metro Toronto, and the Greater Vancouver Regional District have specified responsibilities for provision of some services and for economic planning that avoid wasteful duplication and competition. Another example is Pittsburgh, in which the Allegheny County Commission plans for the metropolitan region of the city and its suburbs.

The need for metrowide governance is given added urgency by Blackley and Greytak (1986) in their study of the location of economic activity within the metropolitan area of Cincinnati. They highlight a pattern of location that is heavily influenced by the ongoing substitution of capital for labor, a need for sites large enough to accommodate capital intensive production, and changes in the technology of transportation and communication. The result is a reduced incentive for firms to cluster in the traditional core and a positive incentive to spread their activities throughout the metropolitan or even regional economic space. Without an effective mechanism for coordinating planning and decision making throughout the relevant area, the individual political units will be led to compete against each other, perhaps giving rise to bidding wars to influence firm locational decisions, a process that Taylor (1992) demonstrates can lead to significant waste of scarce resources.

But it is not just the possibility of waste and duplication that urges effective governance. Grasland (1992) shows in his study of the effort to create a technological center in Montpellier, France, that it was Montpellier as the center of a regional technological space that was important. This involves both physical structures and institutions, as well as budgetary considerations and creation of regional networks that tie the central city together with its outlying towns, universities, and firms.

This section may be summarized by this rather understated comment by Cheshire (1990) in a study of major urban economies in the European Community: "Although most of urban performance seems to be determined by factors over which policy can have no influence, there still remains a small but substantial differential element in comparative urban performance that can be closely related to qualitative information on urban policy" (p. 332).

Urban Strategy

The goal of any urban strategic plan is to mobilize local resources, to create an effective structure for their interaction, to precisely identify an end objective, and to give specific tasks to each of the relevant entities. If this is done well, the city has a chance to achieve the sort of economic future it seeks; if not, the city's economy will have to accept whatever activities come its way. This point was made strongly by Porter (1990), as quoted previously in this chapter. In his work bridging geography and international economics, Krugman (1991) stresses the importance of going beyond the traditional focus on tastes, technology, and factor endowments to include such concepts from economic geography as the spatial location of economic activity, core and periphery, activity clustering, and agglomeration economies to properly explain international trade relations. He concludes, "one needs to think about the geographical structure of production, not treat countries as natural units of analysis" (p. 87).

In their study of four mature industrial regions, Pittsburgh, Nord-Pas de Calais, the Ruhrgebiet, and west central Scotland, Singh and Borzutzky (1988) demonstrate the importance of public-private sector partnership, "the nature of the political organization and the role that different social groups play in the policy process" (pp. 224-225). Given the differing political and social contexts of the four regions, there is no one formula for success, but they stress in general "new approaches as well as new institutions geared to deal with the process of readjustment" and "the appearance of new patterns of interaction between the private and the public sector" (p. 225). Furthermore, they argue that in the fragmented political system of the United States, success may well depend on "the ability of each affected region to influence the central government as well as on its own capacity to adapt to the new domestic and international economic environment" (p. 226).

Shachar and Felsenstein (1992) add the cautionary note that "effective local economic development policy needs to be formulated and executed locally." They give added color to this by noting in their study of urban economic development in Israel that "new high technology companies do not choose locations in the sense of shopping between places. . . . These firms are likely to be born into a location" (p. 853). In their first stage of development, these firms will be interested primarily on survival and markets, and this suggests desirable policies for the local government. Later it is imperative that the city officials work to assist these firms in establishing international linkages. Thus, local governments will have to show flexibility through time and will have to develop supportive policies for firms in different stages of their development.

Heenan (1977) posed the question in his study of global cities why Coral Gables, Florida, was so successful in attracting business decision-making functions for companies operating in Latin America—why not Miami? His response is simply that Coral Gables aggressively sought to develop that role.

> Ironically, the top officials in its Office of Community Development began their careers in Miami, where many of the initial concepts of building a headquarters city were developed. But Miami was unable to secure the necessary support for regionalism. Frustrated, its best city planners moved to Coral Gables and won the support of this smaller, more cohesive community. Subsequently, a nucleus of regional offices was established; Coral Gables has never looked back. (p. 88)

This provides a prime example of Cheshire's (1990) "small but substantial" (p. 332) role for activism on the part of local governments.

Public-Private Cooperation

I begin this section by referring again to the work of Singh and Borzutzky (1988) on mature industrial regions. In commenting on Pittsburgh, they stress the importance of the "re-emergence of a pattern of co-operation between civic and business leaders that has characterized Pittsburgh since the early 20th century." In Nord-Pas de Calais, it was "regional policies geared toward aiding the private enterprises creating a new form of partnership between the regional government and the private sector" (pp. 224-225), and in Glasgow the Regional and District Councils have been active in establishing a variety of public-private sector initiatives and a science park in which the firms and the universities are brought together. Each of these examples was considered to have been relatively successful. In the less successful Ruhrgebiet, there was less of this and what was attempted was of mixed success. Another source mentioned earlier, Shachar and Felsenstein (1992), stresses the importance of local government and its initiatives in the formation of new firms; they warn that if the central government takes on too many functions, the local government is seen by business as little more than a taxing agent and as a detriment to their success and growth.

The basic point here is rather straightforward and generates no controversy. For an effective mobilization of local resources and energies that will enhance a city's competitiveness, both the public and the private sectors must be involved, and the coordination of their activities must be closely monitored. There are, of course, two alternative approaches: (a) a city with no taxes and no services and (b) a city with higher taxes and more services. Each approach will be appropriate to a specific economic structure. Although the former may support low-level assembly and low-value-added manufacture, only the latter will enable a city to develop the production of goods and services that will allow it to move up in the urban hierarchy.

Institutional Flexibility

From the institutionalists, we understand the power of the "dead hand" of past-binding institutions that inhibit the introduction of technology, innovation, and indeed all new ways of doing things. The ossification that results from counterprogressive institutional behavior can be the result of one or more things gone wrong. A bureaucratic structure that is unresponsive to change is the most obvious problem. Another is a politicized process of decision making in which participation is limited to a few insiders.

The element of inflexibility that has gained the most attention from economists is that of excessive regulation, a structure of regulation of activity that (a) makes it exceedingly costly for a firm to comply and (b) imposes constraints on the firm's actions that reduce its flexibility and responsiveness to opportunities. This is, of course, a two-edged sword because although some government regulation burdens participants in market activities, others can reduce uncertainty for firms and thereby enable them to concentrate more productively on the things that really matter to them.

The importance of institutional flexibility has been emphasized in several of the readings referred to throughout this chapter. But because of the impossibility of converting this element of competitiveness into a series of data that can be used in an empirical study, I will have to leave this as something that should be encouraged by the city planner but that cannot be quantified by the researcher.

Conclusions

The primary conclusion of this study of the determinants of the competitiveness of a city's economy is that the individual city's government and private sector entities can do a great deal to enhance that city's competitiveness and to enable it to achieve the most desirable economic future possible. Cities find themselves more exposed to international market and production forces than ever before and more vulnerable to the challenges to existing economic activities, but cities are also more able to take advantage of opportunities for improvement and revitalization than has ever been the case.

The existing research on the individual aspects of city competitiveness provides a rich body of suggestions for optimal policy for a city in the abstract. Because these have been developed previously in this chapter at some length, they will not be reiterated here. But the general conclusion is that cities should stick to the basics—investment in infrastructure and human capital, promotion of smaller firms, ensuring an adequate complement of business and financial service providers, articulation of a well-thought-out and clearly expressed strategic plan, effective governance, and a supportive regulatory environment.

Each city will, of course, be obliged to examine carefully its options, its resources, its intercity competition, and its qualitative strengths and weaknesses when designing its own strategy. This is a challenge that must be met in the increasingly competitive and internationalized economic environment in which individual cities now exist. The cities that fail to address this issue will in all probability be relegated to lower growth of output and employment and to marginality.

NOTE

1. Another approach was offered several years ago by Heenan (1977). This, however, was intended to be used in a survey of business executives and is more qualitative in nature.

REFERENCES

Beeson, P. E. (1990). Sources of the decline of manufacturing in large metropolitan areas. *Journal of Urban Economics, 28,* 78-84.

Beeson, P. E. (1992). Agglomeration economies and productivity growth. In E. S. Mills & J. F. McDonald (Eds.), *Sources of metropolitan growth* (pp. 19-35). New Brunswick, NJ: Rutgers University, Center for Urban Policy Growth.

Beyers, W. B. (1992). Producer services and metropolitan growth and development. In E. S. Mills & J. F. McDonald (Eds.), *Sources of metropolitan growth* (pp. 125-146). New Brunswick, NJ: Rutgers University, Center for Urban Policy Growth.

Blackley, P. R., & Greytak, D. (1986). Comparative advantage and industrial location: An intrametropolitan evaluation. *Urban Studies, 23,* 221-230.

Bonomi, A. (1990). Technological innovation and territorial organization. In Ajuntament de Barcelona (Ed.), *La ciudad ante el 2000* [The city on the threshold of the year 2000] (English trans., pp. 5-7). Barcelona, Spain: Cuaderno Central No. 15.

Calem, P. S., & Carlino, G. A. (1991). Urban agglomeration economies in the presence of technical change. *Journal of Urban Economics, 29,* 82-95.

Cheshire, P. (1990). Explaining the recent performance of the European Community's major urban regions. *Urban Studies, 27,* 311-333.

Clark, D. E., & Cosgrove, J. C. (1991). Amenities versus labor market opportunities: Choosing the optimal distance to move. *Journal of Regional Science, 31*(3), 311-327.

Clark, D. E., & Hunter, W. J. (1992). The impact of economic opportunity, amenities and fiscal factors on age-specific migration rates. *Journal of Regional Science, 32*(3), 349-365.

Dertouzos, M. L., Lester, R. K., & Solow, R. M. (1990). *Made in America.* New York: HarperPerennial. (Original work published 1989)

Florida, R., & Smith, D. F., Jr. (1992). Venture capital's role in economic development: An empirical analysis. In E. S. Mills & J. F. McDonald (Eds.), *Sources of metropolitan growth* (pp. 187-209). New Brunswick, NJ: Rutgers University, Center for Urban Policy Growth.

Fry, E. H., Radebaugh, L. H., & Soldatos, P. (1989). *The new international cities era: The global activities of North American municipal governments.* Provo, UT: Brigham Young University, David M. Kennedy Center for International Studies.

Goldberg, M. A., Helsley, R. W., & Levi, M. D. (1989). The location of international financial activity: An interregional analysis. *Regional Studies, 23*(1), 1-7.

Grasland, L. (1992). The search for an international position in the creation of a regional technological space: The example of Montpellier. *Urban Studies, 29,* 1003-1010.

Hansen, N. (1987). The evolution of the French regional economy and French regional theory. *The Review of Regional Studies, 17,* 5-13.

Harrison, R. T., & Hart, M. (1990). The nature and extent of innovative activity in a peripheral regional economy. *Regional Studies, 24,* 383-393.

Heenan, D. (1977). Global cities of tomorrow. *Harvard Business Review, 55*(3), 79-92.

Heilbrun, J. (1992). Art and culture as central place functions. *Urban Studies, 29*(2), 205-215.

Klaassen, L. (1987). The future of the larger European towns. *Urban Studies, 24,* 251-257.

Kresl, P. K. (1989). Your soul for a case of Coors? Canada-U.S. free trade and Canadian culture policy. In P. K. Kresl (Ed.), *Seen from the south* (pp. 163-168). Provo, UT: Brigham Young University Press.

Kresl, P. K. (1991). Gateway cities: A comparison of North America with the European Community. *Ekistics, 58*(350-351), 351-356.

Kresl, P. K. (1992). *The urban economy and regional trade liberalization.* New York: Praeger.

Kresl, P. K., & Singh, B. (1994, November). *The competitiveness of cities.* Paper presented at the Cities and the Global Economy conference sponsored by the Organization for Economic Cooperation and Development (OECD) and the Commonwealth of Australia, Melbourne, Australia. Forthcoming, OCED, 1995.

Krugman, P. (1991). *Geography and trade.* Cambridge: MIT Press.

Moomaw, R. L., & Williams, M. (1991). Total factor productivity growth in manufacturing: Further evidence from the states. *Journal of Regional Science, 31*(1), 17-34.

Mullen, J. K., & Williams, M. (1990). Explaining total factor productivity differentials in urban manufacturing. *Journal of Urban Economics, 28,* 103-123.

O hUallacháin, B., & Satterthwaite, M. A. (1992). Sectoral growth patterns at the metropolitan level: An evaluation of economic development incentives. *Journal of Urban Economics, 31,* 25-58.

Peirce, N. R. (1993). *Citistates: How urban America can prosper in a competitive world.* Washington, DC: Seven Locks Press.

Pompili, T. (1992). The role of human capital in urban system structure and development: The case of Italy. *Urban Studies, 29,* 905-934.

Porter, M. (1990). *The competitive advantage of nations.* New York: Free Press.

Scott, B. R., & Lodge, G. C. (1985). Introduction. In B. R. Scott & G. C. Lodge (Eds.), *U.S. competitiveness in the world economy* (pp. 1-12). Cambridge, MA: Harvard Business School.

Shachar, A., & Felsenstein. (1992). Urban economic development and high technology industry. *Urban Studies, 29,* 839-855.

Singh, A. (1977). UK industry and the world economy: A case of de-industrialization? *Cambridge Journal of Economics, 1*(2), 113-136.

Singh, V. P., & Borzutzky, S. (1988). The state of the mature industrial regions in Western Europe and North America. *Urban Studies, 25,* 212-227.

Soldatos, P. (1991). Strategic cities alliances: An added value to the innovative making of an international city. *Ekistics, 58*(350-351), 346-350.

Taylor, L. (1992). Infrastructural competition among jurisdictions. *Journal of Public Economics, 49,* 241-259.

Van den Berg, L., & Klaassen, L. H. (1989). The major European cities: Underway to 1992. In M. Belil (Ed.), *Eurocities* (pp. 55-60). Barcelona, Spain: Organizing Committee of the Eurocities Conference.

3 Cities and Citizens in Flux: Global Consumer Integration and Local Civic Fragmentation

JOHN KINCAID

The contemporary world is being shaped and reshaped by many forces having positive and negative effects on cities. As Hoffman (1966) suggests, "Every international system owes its inner logic and its unfolding to the diversity of domestic determinants, geohistorical situations, and outside aims among its units" (p. 864). Two overriding determinants of the late 20th century are global economic integration and the end of the Cold War. These forces have produced or unleashed somewhat contradictory trends having significant implications for cities, including (a) variable integration of world regions, (b) national fragmentation, (c) intranational decentralization, (d) tensions between local citizenship and global consumership, and (e) disruptions of historic connections between peoples, places, and persons. Under these conditions, some cities prosper, whereas others suffer the fates of Sarajevo and Kigali (Rwanda). As a result, "the quality of urban life around the globe varies tremendously" (Gappert, 1989, p. 306) and will vary even more if current trends persist.

■ Variable Integration

Although much attention has been given to the benefits of free trade and global economic integration, the reality is that of the world's nearly $20 trillion gross national product (GNP) in 1991, the European Community (EC) accounted for $6.5 (U.S.) trillion, the United States for $5.5 trillion, and Japan for $3.4 trillion. Measured in GNP-equivalent terms, moreover, corporations constituted 57 of the world's leading 100 economic powers.

If future growth tracks past trends, no other region or nation will overtake the EC, the United States, or Japan during the next two decades.

One result of this reality is that efforts to define global cities as universal phenomena encounter difficulties because most of the world's cities and their citizens operate at the margins of the global economy. If a global city is said to be knowledge oriented, science based, communicative, high-tech aspiring, culturally open, driven by market forces, and engaged in a range of international activities (e.g., Knight, 1989), then most cities fail to meet most of the criteria.

Worst off is urban Africa, home of 18 of the world's 25 poorest countries. Excluding South Africa, the total gross domestic product (GDP) of sub-Saharan Africa, with a population in excess of 450 million, is no larger than the GDP of Belgium, which has some 11 million people. Altogether, Africa produces only about 1% of the world's GDP.

The immediate prospects are not promising and may not improve as long as dictatorial regimes remain intact and ethnic conflict, such as the 1994 apocalypse in Rwanda, fragments governments and nations. Many nationalist leaders who emerged from the anticolonial movements valued political control and centralized unitarism more than political freedom and democratic pluralism. Although African nationalists "were among the leaders in developing the post-1945 'right' to self-determination" (Hannum, 1990, p. 46) of all peoples against colonization, most of them rejected ethnic self-determination and embraced nation-state self-determination and national unification of ethnic heterogeneity. For these and other reasons, attempts to establish viable Pan-African institutions, including federal arrangements and common markets, have not been notably successful.

Perhaps new leaders emerging from postcolonial generations, for whom domestic repression rather than foreign oppression has been the dominant experience, may support political freedoms and pluralist democracy, but collapsing dictatorial regimes may unleash many centrifugal forces propelled by ethnic, linguistic, and religious autonomy claims. As a result, cities as much as rural areas are potential killing fields, and the absence of competent national governance, needed for urban economic development, may drive many African cities into further misery.

Although the oil-rich Middle East is integrated into the world economy, Islamic resistance to Westernization poses barriers to broad economic, social, and political integration—as reflected, for example, in the widespread ownership of automobiles that cannot be driven by women in some countries. Only Lebanon and Turkey officially disestablished Islam, although Lebanon ceased to function as a viable nation-state in the 1980s, and Beirut is the leading example of an emerging global city destroyed by ethnoreligious violence. Otherwise, except for Israel, Middle Eastern nations lacking valuable natural resources are not well integrated into the global economy or society.

Islamic societies illustrate the possibility of some modern cities having many of the technological accoutrements of global cities while remaining defensive and insulated from global society and the cultural values associated with it. Islamic perspectives expose the underlying nature of the global economy, namely, Westernization. Most of the characteristics commonly attributed to global cities are rooted in Western, often American, values; consequently, building global cities is not, in the eyes of some cultural elites, an improvement of civilization. This is especially true in modern communications, in which international media regulation is highly controversial and resistant to free trade because many national elites wish to maintain cultural walls against Western media invasions.

Much of non-Muslim Asia, however, appears to be more open to globalization. The editor of the Bangalore edition of *The Times* of India, for example, has suggested that the traditions of the British empire are still prominent in Bangalore, India's Silicon Valley. The city embraces the West and seeks economic development (Anderson, 1993). Of the world's five nations having the most purchasing power in 1990—the United States, China, Japan, Germany, and India—three are located in Asia. Two of the region's great nations, moreover—Japan and Australia—are full participants in global integration. China, which experienced a 9.5% average annual GDP growth rate from 1980 to 1990, is integrating into the global economy, although less so into the global society.

A major characteristic of Asia's economic "tigers" has been export promotion, which has stimulated the manufacturing and technological developments needed for economic growth. To produce exports, these countries aggressively acquired the foreign skills required for modern manufacturing and studied foreign consumer preferences and market practices to sell goods abroad. The Japanese, of course, successfully penetrated the American consumer market, even rivaling the once world-dominant U.S. automotive industry. Japan's share of the U.S. automobile market reached 30% in 1991, then dropped to 23% in 1994.

Consequently, Asia's ancient urban civilizations have been recovering from colonialism or foreign occupation and building some of the world's great modern global cities, such as Tokyo, Singapore, and Hong Kong, with more cities likely to reach global status in the near future. At the same time, Asia has not thoroughly capitulated to Western globalization, in part because the ability to export also creates the ability to produce for domestic consumption and reduce dependence on foreign imports. In addition, producing globally competitive goods can be a source of cultural pride.

Much of historically poor Latin America seems poised to enter the global economy as a significant player. Recent movements toward democratization and pragmatic regional integration seem to be providing more favorable bases for global integration. Argentina and Brazil signed a *Programa de Integración Argentino-Brasilena* in 1986, to which Paraguay and Uruguay signed on in 1991 in the Treaty of Asunción. Colombia, Mexico, and Venezuela intend to create a common market. Efforts also are under way to revive the Central American Common Market consisting of Costa Rica, El Salvador, Guatemala, Honduras, Nicaragua, and Panama. Mexico, which had one of the region's most dynamic economies until the 1995 peso crisis, is a force for integration, and some of the region's integrating activity reflects an expectation that the North American Free Trade Agreement (NAFTA) can be extended to Latin America. Of course, through drug exports, some Latin American countries are well integrated into the underside of the global economy, with devastating effects on life in many cities in Latin America and abroad.

Most of the new nations spawned by the collapse of the Soviet Union seem eager to embrace economic integration, but there is tremendous variation in the ability of these nations and their cities to engage the world economy effectively. Even Moscow, which was arguably a global city when the Soviet Union was a superpower, is not likely to emerge as a modern global city in the near future; instead, such Eastern European cities as Prague and Warsaw, as well as Berlin, appear likely to be the first global cities to emerge from the former Soviet sphere of power, in part under the tutelage of the European Union (EU) and the North Atlantic Treaty Organization (NATO).

Outside of Japan, therefore, the most dynamic centers of global integration are North America (Canada, the United States, and, increasingly, Mexico) and Western Europe. These two regions constitute the best organized trade blocs—NAFTA and the EU—and their major cities most closely fit the criteria of global cities. Yet, even many of these nations and cities are cautious about globalization because of such problems as trade deficits, in-migration from other regions, and the culturally corrosive effects of global integration. Western Europe, especially France, for example, insisted on trade barriers against American popular culture exports in the Uruguay Round of the General Agreement on Tariffs and Trade (GATT) negotiations; Americans and Canadians have expressed grave concerns about trade deficits and losses of their manufacturing bases to foreign competitors; and both Western Europe and North America have been politically agitated by foreign immigration.

Thus, global economic integration is highly variable; it is not embraced enthusiastically by all cultures; and even the regional leaders of integration remain economically ambivalent and culturally resistant to rapid political and social integration. Although major cities are the nerve centers of global integration, they are parts of larger body politics, which have authoritative responsibility for the national and international laws that regulate or fail to regulate the economic forces affecting cities (Kincaid, 1990).

National Fragmentation

The new world order, however, is also marked by increased fragmentation of nation-states. The possible breakup of Belgium and Canada; regional fracturing in Italy and Spain; secession and warfare in the Balkans; the disintegration of Lebanon, Rwanda, Somalia, and Yugoslavia; assertions of ethnic nationalism, regional consciousness, and religious revivalism; regional and local government forays into foreign affairs; large-scale legal and illegal movements of people across borders; and other forces having centrifugal effects on nation-states all seem to point to an apparent paradox of national disintegration in the face of global integration.

A major factor in national fragmentation, however, has been the collapse of the Soviet Union and, therefore, the falling of authoritarian regimes that had imposed unity on many heterogeneous societies. Ethnic and religious competition and conflict have reappeared in the power vacuums and instability created by regime changes and by greater exposure to global economic and social influences. Opportunities have arisen for nationalist leaders to gain power and legitimacy among groups having long-suppressed resentments, particularly groups that have regarded themselves as captive nations. Thus, a critical question that has long plagued the nation-state epoch—whether every nation (i.e., people or ethnic group) can or should have its own state—has resurfaced in the more open climate of the post-Cold War era.

A corollary following from the collapse of the Soviet Union has been the decline of the "hegemonic stability" (Keohane, 1984, p. 11) of the Cold War's bipolar rivalry and, therefore, of a willingness of the major powers to guarantee the national unity of former client states. Global communications allowed the world to watch the violent disintegration of Yugoslavia,

Somalia, Rwanda, and Lebanon on television, but the world powers and leading forces for global integration were unwilling to halt the disintegration. Given that the formation of nation-states reduced many nations to the status of ethnic groups and given the weakness of international mechanisms for coping with civil conflicts, incentives for national disintegration are likely to proliferate in many regions.

These conflicts have had deleterious consequences and, in some cases, violent results for cities and their citizens. The nation-states and urban areas subject to violent civil conflict have generally lacked the pluralist democratic institutions and intergroup processes needed to accommodate human diversity. Furthermore, the emergence of ethnic cleansing reveals the terrible extent to which some national governments and cities in heterogeneous societies have no desire to accommodate diversity. In these societies, various cities lay claim to being national (i.e., ethnic-group) capitals having rights to the same international status as other national capitals. Hence, urban elites in ethnically heterogeneous nation-states whose cities are competitively disadvantaged by the existing regime's national capital and cultural centers have strong incentives to agitate for secession or substantial home rule. Likewise, urban elites in more affluent regions may seek to jettison less affluent regions (e.g., southern Italy and southern Mexico) that retard economic development and require subsidies. For example, the Czech Republic and its emerging global city, Prague, had few cultural or economic reasons to remain tied to Slovakia. Thus, both more and less affluent regions in heterogeneous societies have varying incentives to dissolve national political bonds.

Global economic integration under conditions of free trade also contributes to national fragmentation insofar as reduced national authority to regulate and maintain trade barriers exposes intranational regions and their cities to more unrestrained economic competition. This may be especially salient in regional trade blocs. Under the umbrella of the EU, for instance, it becomes politically possible and more economically rational for certain subnational regions, such as those in Belgium, Germany, Italy, and Spain, to agitate for more autonomy and to consider participating as direct partners in the EU rather than as stepchildren of their nation-states. Indeed, the possibility of a Europe of the Regions is no longer unthinkable. Many cities in Western Europe have been linking with regional neighbors across borders, and the German *Länder* (i.e., states), for example, have been establishing Euroregions for cross-border cooperation. Similarly, under NAFTA, it becomes easier for Quebec to ask whether it really needs to remain a part of the rest of Canada. A key question, though, is whether Montreal could maintain its global-city status in an independent Quebec. Thus, actual and potential national fragmentation, especially in Africa, Eastern Europe, and Euro-Asia, has created a highly fluid and uncertain political and economic environment for many cities, which must evaluate not only their international position in the global economy but also their regional position in their national economy.

Intranational Decentralization

Short of actual fragmentation, the more prevalent phenomena under conditions of global integration and democratization are likely to be intranational decentralization and devolution. For example, with respect to Spain, what will Madrid do once it has delegated half its powers to Brussels and devolved the other half to the regions? One can conceive of a range of diminished nation-state autonomies and national government powers induced by global integration and the end of the Cold War.

First, in some heterogeneous societies reacting against the poor economic and human-rights performance of centralized, authoritarian regimes, the high price of civil war may stimulate nonviolent accommodations of diversity through political decentralization, federalism, home rule, and regional representation arrangements. Because the heterogeneous nation-states most vulnerable to civil war are already weak competitors in the global economy, destruction of their urban infrastructure would condemn them to decades or more of deeper poverty.

Second, the market-based character of contemporary integration requires divestment of certain powers historically exercised by national governments. Economic liberalization also tends to produce pressures for political decentralization and democratization. These developments may undermine the legitimacy of certain regimes but not necessarily of the nation-state itself, thus producing decentralization rather than fragmentation.

Third, increased global integration and decreased military threats have demonstrated that small nation-states, quasi-states, and even city-states can be extraordinarily viable economically. On the territorial scale of state organization, Taiwan, Hong Kong, Singapore, Japan, South Korea, Kuwait, Israel, and Switzerland, for example, are tiny, and most are not well endowed with natural resources. Being a large resource-rich nation-state, therefore, is not itself a competitive advantage. Consequently, it is not necessarily irrational for subnational regions and major cities to assume that they might be better off as independent or semiautonomous states.

Fourth, international economic integration tends to heighten the importance of subnational regional economies and accentuate differences between those economies. The most relevant territorial competitors in the global economy are specific urban regions able to capture investment, attract tourists, and produce exports. The diminished authority of nation-states that wish to participate in the global economy to control the flow of capital and commerce across their borders under rules of free trade also diminishes their ability to function as traffic cops or central planners directing capital and commerce to specified regions and cities. Hence, urban regional competition within nation-states can be accentuated, and cities that view their success in the global economy as due as much or more to their own entrepreneurial policies as to their nation-state's policies may be reluctant to support national policies that redistribute their profits to laggard regions, especially through what may be viewed as a bloated national bureaucracy (e.g., Moscow, Mexico City, and Washington, D.C.).

Hence, the end of the Cold War and the rise of global economic competition are creating a type of free market of governments, not only of national governments but also of regional and local governments. These latter governments, especially, are becoming more entrepreneurial and are entering the global marketplace directly to recruit investment and tourists and to promote exports (Brown & Fry, 1993; Fry, Radebaugh, & Soldatos, 1989). More regional and local governments also are seeking voices in the decision making of national and international institutions (e.g., German *Länder* representatives in Brussels) that shape the rules of trade and economic life. Under the pressure of competition, therefore, regions and cities may increasingly seek to be deregulated by their national governments.

At the same time, there is no rush to dismantle the nation-state system, and many regions and cities are not eager to abandon the protections they receive from their national governments. Global integration is highly pragmatic and quite segmented functionally and regionally. Aside from the vague term *globalism,* no overriding ideology or "ism" drives integration. The process is largely one of pragmatic cost/benefit decision making by business enterprises and governments. Integration is proceeding incrementally, at different speeds, and in different ways in various functional areas and geographic regions. Quasi-ideologies shape some functional areas (e.g., environmentalism), and ideologies influence some regions (e.g., Islamic fundamentalism), but for the most part, developed nation-states are at the helm of integration, endeavoring to respond to citizen interests in life's more mundane "isms," especially consumerism, tourism, and television.

Citizenship Versus Consumership

Indeed, the global spreading of consumerism and the desire of ever more citizens to gain access to the global marketplace are producing significant tensions between historic rights of citizenship and the attractions of modern consumership (Kincaid, 1993). At present, this tension is most evident in the developed liberal democracies in which consumption and individual rights, rather than communal identity and civic duties, are increasingly central to personal autonomy and self-actualization. Although most citizens still shop for groceries at their hometown center, the global economy is increasingly the relevant marketplace. Standardized products, such as Coca-Cola and McDonald's, are available almost everywhere, and specialized products from all parts of the world are available to ordinary citizens in the developed world and to elites in the less-developed world. Even where products are not directly accessible, telecommunication advertises global goods to citizens in many poor and remote villages. The advent of telecommunication shopping, moreover, allows citizens to transcend borders by entering the global market from their home or village telephone booth. At the same time, however, as citizens of specific places, consumers often wish to exercise powers of municipal and national self-government that may not conform to the rules of global integration. Thus, the more that citizens wish to consume global goods, the more they must consent to international preemptions (i.e., displacements) of their powers of local, regional, and national self-government.

Integration allows, and advertising encourages, citizens to assert rights to acquire the goods and services sold in the global market, whether consumed at home or consumed abroad through tourism. Yet maintenance of a global market characterized by free trade and efficient resource allocations entails limits on local and national self-governance and, thereby, restrictions on the exercise of historic citizenship rights. This tension is felt by both persons and governments wishing to foster a global marketplace able to satisfy consumership interests while still maintaining rights of local self-government able to satisfy citizen interests in local economies, social policies, cultural preferences, and political choices. Most persons wish to be citizens of an identifiable place, ordinarily a state or community infused with primordial sentiments, but increasingly they also desire the means and freedom to consume the goods and services available on the world market. Yet the more that persons desire global consumership, the more they emphasize individuality as defined by their behavior as autonomous consumers and the more they discount their duties as loyal citizens of a nation-state or city. Furthermore, the term *global city* itself connotes the possibility of conflict between local citizens' global and municipal interests.

The evolution of shopping in the United States offers a mundane but telling example of this tension. Despite repeated expressions by Americans of their love for hometown values and local community, consumers abandoned their downtowns and Main Streets for the suburban shopping malls that proliferated after World War II when the United States developed a truly integrated national consumer economy rather than simply a manufacturing economy. Few Americans regard local shopping as a local civic duty, yet they continue to defend local government against big government. Similarly, as citizens, individuals may be boosters of their state of residence, but as consumers, they rarely hesitate to cross state lines to purchase goods at a lower sales tax rate or to purchase out-of-state goods by mail to escape state and local sales taxation. Hence, mail-order selling is a rapidly growing industry that costs state and local governments about $3.9 billion in lost tax revenue in 1992 (Coleman, 1992).

Global incursions on local citizenship powers in order to enhance consumership interests are accelerating. Implementation by the 119 participating nation-states of the Uruguay Round of GATT and the General Agreement on Trade in Services (GATS), completed in December 1993, carries unprecedented implications for regional and local self-government, especially in federal systems such as the United States, in which states are cosovereign polities and in which county and municipal governments have substantial autonomy. GATT and GATS will allow, and sometimes require, new federal government regulation and preemption of state and local government powers in a wide range of fields, including taxation, spending, economic development, environmental protection, procurement, land use, and surface transportation as well as state and local regulation of general services, professional occupations, securities, banking, insurance, other financial services, and health and safety (Weiler, 1994).

Many state laws and municipal ordinances are potentially subject to challenge by foreign governments. Indeed, the EU has already prepared a hit list of offensive state laws and local ordinances. Foreign enterprises especially desire access to the $200 billion state and local government procurement market in the United States. Furthermore, for example, cities and states will be prohibited from offering certain types of subsidies to attract or retain businesses. GATS may also constrain the ability of state and municipal governments to provide tax breaks to nonprofit providers of public services and for interest accrued on municipal bonds. Thus, key aspects of city and metropolitan governance will come under international constraints.

State and local governments will also be subject to rulings by international bodies, particularly the new World Trade Organization (WTO). Proposed legislation in Congress to implement GATT will authorize the U.S. Trade Representative (USTR) and the U.S. Department of Justice to preempt state and local laws pursuant to WTO concerns and to sue state and local governments to require compliance with the USTR's or WTO's interpretations of GATT. If the legislation is enacted without amendments affording more protections for state and local powers, then state and local governments and their citizens will face two powerful challenges to their rights of self-government.

For one, the USTR—an executive-branch agency—will acquire unprecedented power to preempt state laws and municipal ordinances. The USTR will also be able to block proposed bills in state legislatures and proposed ordinances in county commissions and city councils deemed to be in conflict with GATT or GATS. No executive-branch agency has ever possessed such powers of preemption in the American federal system. At the same time, the USTR will not possess this power over federal law. If the USTR deems a federal law to be in conflict with GATT or GATS, the USTR will be required to ask Congress to change the law, which the Congress may refuse to do under its own powers to interpret international agreements. Given that states and localities have representatives elected to Congress but not to the USTR's office, many state and local officials argue that the USTR should be required to refer GATT-conflicting state and local laws to Congress for preemption decisions. This would afford greater protection for state and local laws; however, Congress itself has already preempted more state and local laws since 1969 than it did during the entire previous 180 years (1789-1969) of U.S. history (U.S. Advisory Commission on Intergovernmental Relations, 1992).

Second, foreign corporations will be able to lobby the USTR to declare state laws and municipal ordinances in violation of GATT or GATS so that these firms can litigate against states and localities. Similarly, the WTO's preemption decisions will be enforceable against state and local governments by foreign corporations in U.S. courts by using WTO decisions and USTR opinions as evidence in cases brought under the U.S. Constitution's foreign-commerce clause (Article I, Section 8). In addition, although U.S. corporations are not direct beneficiaries of USTR enforcement within the United States, large U.S. corporations may piggyback equal treatment cases on to lawsuits filed by foreign corporations. State and municipal tax systems and procurement practices are likely to be particular objects of foreign corporate legal challenges.

Furthermore, through GATT as well as NAFTA, U.S. corporations can indirectly precipitate international preemptions of state laws and municipal ordinances—preemptions not otherwise obtainable through the U.S. Congress or courts. For example, U.S. brewers initiated a GATT action against allegedly discriminatory Canadian provincial beer and wine laws. In *Beer I,* a GATT panel ruled against the provinces in 1988. In retaliation, Canada initiated a GATT investigation of U.S. federal, state, and local laws governing beer and wine, including even Mississippi's local-option law. In *Beer II,* a GATT panel ruled favorably in 1992 on most of Canada's complaints. For example, the panel ruled against a Minnesota law that gave a tax preference to beer from microbreweries, even though this preference was given to microbreweries anywhere in the world (i.e., equal tax treatment). The panel argued that the tax preference was unrelated to the product, namely, beer, and therefore discriminated against large foreign producers of beer. Hence, large U.S. brewers also gained significant tax victories from both *Beer I* and *Beer II.*

An example of another significant implication of *Beer II* for cities is that a municipal government that contracts with a tax-exempt charitable organization to provide a public service might be challenged by a foreign for-profit corporation selling the same service. The foreign corporation may claim tax discrimination under GATT and thus demand the same tax exemption or else equal taxation of the domestic charitable organization.

Although states and municipalities have won many cases challenging their laws, their freedom of action is increasingly constrained by international agreements. They must now be cognizant of these agreements in framing laws, and they must be prepared and financially able to engage in defensive litigation. Trade between the United States and Canada, for example, continued to increase at about the same rate after the Canada-United States Free Trade Agreement as before the agreement, but dispute-resolution cases shot up sharply under the agreement.

Although *Beer II* served to crystallize state and local government fears of the preemption implications of free trade agreements, most governors, state legislatures, and mayors support NAFTA and GATT. Mexico's inclusion in NAFTA, for example, was supported by 41 governors in 1993. Only seven state legislatures passed resolutions opposing NAFTA. The principal opposition came from state and local officials in the Rust Belt where there were widespread fears of manufacturing losses to Mexico.

State and local government officials are caught in the same tension between citizenship and consumership. Although they endeavor to preserve the powers of self-government they are elected to exercise, they must nevertheless satisfy the consumership demands of their constituents. Most U.S. state and local officials have concluded that the economic and revenue benefits of such agreements as NAFTA and GATT outweigh the costs to federalism and local self-government. States having existing or emerging global cities, and these cities themselves, must especially surrender certain rights and powers of their citizens in order to be effective players in the global marketplace.

These developments highlight another seemingly contradictory trend in global integration. On one hand, international agreements and global economic forces reduce the power of national governments by prohibiting or inhibiting their regulation and taxation of the movement of capital, labor, goods, and services across frontiers and within the nation. On the other hand, national governments are called on to exercise certain powers as guarantors of national and subnational compliance with international rules. For example, nation-states still have the primary role in implementing and enforcing EU rules, and the Council of Ministers remains the EU's principal decision-making power, although the European Court has curbed national sovereignty in certain policy areas.

The world's federal nation-states face some of the most difficult political challenges in adjusting to global integration. It is precisely the role of national governments in negotiating and guaranteeing compliance with international agreements that endangers powers historically exercised by the constituent regional and municipal governments of federations. International forums, moreover, even in the EU, are not democratic in the customary sense. They are constituted by national governments, not by representatives directly elected by citizens. How, for example, would one accommodate the Swiss federation's semidirect referenda democracy, with its requirements for popular national and cantonal majorities, to the governing structure of the EU, or even to a reformed structure based on a fully empowered and directly elected parliament? These are among the reasons why Swiss voters, thus far, have elected to remain outside the European economic space. The centralization of power in American federalism induced by global integration may serve as a further object lesson for the Swiss.

Hence, although market-based global integration under conditions of free trade encourages economic deregulation and political decentralization and increasingly requires regions and cities to fend for themselves against global competition, the rules needed to maintain this global integration also require national centralization for enforcement and, thereby, the subversion of regional and municipal self-government. In turn, this may exacerbate civil conflict in ethnically heterogeneous societies in which the national government may be seen as discriminating against certain groups in its enforcement of international rules.

■ Peoples, Persons, and Places

A basic, long-term challenge is global integration's gradual erosion of the historic connections between peoples, places, and persons. Historically, peoples (racial, ethnic, religious, and/or linguistic groups) have been associated with specific territorial places, some of which became great cities and nation-states, and persons have been part and parcel of those peoples and places. Identity for individuals was a composite of people and place, namely, a communal rather than unique individual identity. This is still true in most of the world, but the global spreading of democratization, human rights conventions, market economies, and consumerism is gradually liberating persons from both peoples and places and creating identity opportunities based on personal achievements and lifestyle choices.

Given the accelerating mobility of capital, goods, and services across borders, the next frontier of global integration is human mobility. A genuinely free global marketplace would allow unfettered mobility of persons, much like the mobility that occurs between states and localities in the United States. Despite the already considerable migration in the world, in most nation-states there continues to be strong cultural resistance to extending citizenship to foreigners. Although the concept of a global city implies diversity and receptivity to new citizens from around the world, few global cities are open to such diversity, and nation-states have continued to monopolize citizenship.

The Maastricht Treaty (Part Two), for example, establishes the concept of European citizenship. Aside from relaxed passport controls, however, the concept still has little content. Unlike the United States and most other immigrant polities, which base citizenship on *jus soli,* most EU member-states base citizenship on *jus sanguinis.* Hence, the free migration essential to a common market and political union is made costly by restrictions on the rights of resident aliens necessitated by *jus sanguinis.* As a Commission of the European Communities (1988) report complained, the nearly 5 million EC citizens who have migrated from their home states to other member states "are deprived of the right to vote in local elections simply because they are no longer in their Member State of nationality" (p. 26). During the French referendum on Maastricht in 1992, some voters were appalled by the prospect that a non-French EU citizen could be elected mayor of a French city. Yet can Paris be regarded as a truly global city if a lifetime German or Italian resident does not have an equal opportunity to be elected mayor? Thus, the rule of *jus sanguinis* impedes integration, even in liberal Western Europe.

Short of actual mobility, however, global integration is highly corrosive of the cultural glue that has held the identities of persons to the peoples and places of their birth. Although citizenship may not be highly mobile, consumership is highly mobile as persons circle the globe in search of consumer goods, tourist havens, and investment opportunities. Even cultures and languages, like automobiles, are increasingly matters of consumer choice in the global economy. As was said of returning American veterans after World War I, how can we keep them down on the farm once they have seen Paris? Similarly, persons throughout the world are at least aware of, even if they lack full access to, the global consumer culture. Automobiles, rock and roll, blue jeans, and the like have invaded virtually every cultural space in the world, ungluing those cultures and disrupting traditional patterns of civil behavior.

In this respect, global integration also poses problems of civic fragmentation in cities. This experience is already evident in the case of highly mobile corporations that have few or only lukewarm civic commitments to their current locations. Corporations often extract concessions from cities as the price of locational choices, and cities that fail or refuse to meet corporate needs experience business losses. Global competition and mobility exacerbate these patterns of behavior. Thus, although business is a key element of a city's civic infrastructure, the city itself may be a disposable item for its major businesses. In turn, corporate mobility disrupts residential stability and further weakens the civic infrastructure by requiring citizens to be mobile or by leaving them behind in less favorable economic circumstances.

As more citizens themselves, however, see their fortunes tied to the global economy, they may also, like corporations, have fewer incentives to invest in the civic life of their resident cities. Their interest in the vitality of their city may extend only so far as it benefits their global interests. This is already true of many corporate, academic, and nonprofit sector professionals whose global career interests motivate them to move from place to place in pursuit of career advancement, which takes precedence over notions of local civic responsibility. Similarly, even as immobile citizens acquire more consumer goods, such as automobiles and televisions, they are likely to have fewer incentives to participate in traditional modes of civic life and more incentives to demand changes in city life to accommodate their consumer interests. Furthermore, given the preemptions of municipal powers occurring under the rules of global integration, citizens will have less reason to exert energy to exercise diminished rights of municipal self-government.

Thus, although most cities are likely to retain a core group of civically committed citizens, all cities face major challenges of highly fluid civic, cultural, social, political, and economic environments in the global economy. The nature and extent of these challenges for particular cities vary, of course, with the nature and extent of their integration into the global economy, but the long-run future appears to be toward more integration. This integration will necessitate more international and intercity cooperation and will likely, in the long term, be conducive to prosperity. But along the way there will be considerable conflict, and certain regions exposed to the global consumer culture will remain stagnant at the margins of the global marketplace for the foreseeable future.

REFERENCES

Anderson, J. W. (1993, August 1). Indians, foreigners build Silicon Valley in Bangalore. *Washington Post,* p. A21.

Brown, D. M., & Fry, E. H. (1993). *States and provinces in the international economy.* Berkeley: University of California, Institute of Governmental Studies.

Coleman, H. A. (1992). Taxation of interstate mail-order sales. *Intergovernmental Perspective, 18*(1), 9-13.

Commission of the European Communities. (1988, February). A people's Europe: Proposal for a council directive on voting rights for community nationals in local elections in their member states of residence [Suppl.]. *Bulletin of the European Communities.*

Fry, E. H., Radebaugh, L. H., & Soldatos, P. (1989). *The new international cities era: The global activities of North American municipal governments.* Provo, UT: Brigham Young University, David M. Kennedy Center for International Studies.

Gappert, G. (1989). Concluding perspectives: Thinking globally, acting locally. In R. V. Knight & G. Gappert (Eds.), *Cities in a global society* (Urban Affairs Annual Review, Vol. 35, pp. 301-304). Newbury Park, CA: Sage.

Hannum, H. (1990). *Autonomy, sovereignty, and self-determination: The accommodation of conflicting rights.* Philadelphia: University of Pennsylvania Press.

Hoffman, S. (1966). Obstinate or obsolete? The fate of the nation-state and the case of Western Europe. *Daedalus 95,* 861-887.

Keohane, R. O. (1984). *After hegemony: Cooperation and discord in the world's political economy.* Princeton, NJ: Princeton University Press.

Kincaid, J. (1990). Constituent diplomacy in federal polities and the nation-state: Conflict and co-operation. In H. J. Michelmann & P. Soldatos (Eds.), *Federalism and international relations: The role of subnational units* (pp. 54-75). Oxford, UK: Clarendon.

Kincaid, J. (1993). Consumership versus citizenship: Is there wiggle room for local regulation in the global economy? In B. Hocking (Ed.), *Foreign relations and federal states* (pp. 22-47). London: Leicester University Press.

Knight, R. V. (1989). Introduction: Redefining cities. In R. V. Knight & G. Gappert (Eds.), *Cities in a global society* (Urban Affairs Annual Review, Vol. 35, pp. 15-20). Newbury Park, CA: Sage.

U.S. Advisory Commission on Intergovernmental Relations. (1992). *Federal statutory preemption of state and local authority.* Washington, DC: Author.

Weiler, C. (1994). Foreign-trade agreements: A new federal partner? *Publius: The Journal of Federalism, 23*(3), 113-133.

Part II

The Role of Internationalization in Revitalizing North American Municipalities

4 Globalization and Wage Polarization in U.S. and Canadian Cities: Does Public Policy Make a Difference?

MARC V. LEVINE

Global economic integration, reflected in expanding international trade and capital flows, is now routinely recognized as a basic force shaping the fortunes of cities across North America. Consider trade: In the United States between 1960 and 1993, the ratio of exports and imports to gross domestic product (GDP; the most commonly used measure of the importance of trade in the economy) grew from 9.4% to 21.9% (*Economic Report of the President,* 1994, p. 269). Since 1984, direct investment by U.S. companies in overseas operations has doubled, reaching $716.2 billion in 1993 (Uchitelle, 1994, p. 1); conversely, by 1989, 3.8% of all U.S. workers were employed by U.S. affiliates of foreign companies—up from 1.2% in 1974 (Commission on the Future of Worker-Management Relations [Dunlop Commission], 1994, p. 3).

There are, however, two different perspectives on how these changes in the world economy affect urban life and on how cities should respond to them. One analytic framework, the "global village" perspective, treats global economic integration rather benignly, as a step toward a bright future of a world as "one big marketplace." From this perspective, cities that position themselves as international or competitive will reap economic benefits in the form of investment, jobs, rising incomes, and improved living standards for its residents. Increasingly, adapting to global realities has become a watchword of urban economic development. International trade offices, incentives for foreign investment, and urban redevelopment programs designed to attract corporate headquarters are now standard elements of urban economic development practice in the United States and Canada (Gudell, 1994; Peirce, 1993).

By contrast, a second perspective emphasizes a less than salutary impact of laissez-faire globalization on urban economies. Far from uniformly elevating urban living standards, argue some scholars, unregulated globalization results in bifurcated labor markets and social dualism. Global cities, argues Sassen (1991), "contain both the most vigorous economic sectors and the sharpest income polarization" (p. 337). By the same token, Reich (1991) maintains that global competition has splintered urban labor markets in the United States into two groups: high-earning "symbolic analysts," whose problem-solving skills are in demand around the world and "for whom globalization creates a bright economic future" (p. 282), and the vast majority of production workers and low-wage service providers whose earnings are depressed by global economic integration. Finally, economists such as Harrison and Bluestone (1988; Harrison, 1994) argue that "lean and mean" corporate strategies adopted since the 1970s to cope with heightened global competition—downsizing, outsourcing, multisite production, and relentless cost cutting—have resulted in stagnant real wages in the United States and rising inequality. These polarization tendencies are exacerbated as cities adopt "low road" competitiveness strategies, seeking to attract footloose investment by lowering the costs of labor and social benefits. Thus, in this second framework, unregulated globalization is a fundamental cause of labor market dualism, polarized urban social structures, and declining living standards.

In short, major unresolved questions remain regarding the effects of globalization on North American urban labor markets. Are cities in advanced societies that are more integrated into the global economy (i.e., competitive cities) likely to have more broadly rising living standards than other urban centers? Does the way in which competitiveness is achieved influence urban labor markets? On the other hand, to what extent, as analysts such as Sassen, Reich, Bluestone, and Harrison argue, is unrestrained globalization associated with the polarization of wages and, in particular, the proliferation of low-wage jobs? Finally, do differing public policies, social institutions, and labor market rules mediate the impact of globalization on urban labor markets, resulting in different patterns of earnings and inequality?

A comparative analysis of wage trends in selected U.S. and Canadian cities offers an excellent "natural laboratory" in which to explore these issues. Canadian and American cities inhabit geographically proximate, increasingly interconnected economic space; thus, they are likely to be similarly influenced by trends such as global economic integration and new production technologies that may affect earnings. These structural similarities, as Card and Freeman (1994) have recently pointed out, make it easier to identify the degree to which differences in labor market outcomes in the two countries are attributable to differences in specific institutions or policies or to what extent globalization has a relatively immutable impact on wage patterns in the urban centers of advanced societies. Compared with U.S. cities, the labor markets in Canadian cities feature higher rates of unionization, higher minimum wages, and a more generous array of social benefits available to workers. Have these differences shaped the impact of globalization on the distribution of earnings in cities in the two countries? That will be the central question examined in this chapter.

Globalization and Wage Polarization: The Argument

American economists now agree that income inequality and polarization in the distribution of wages in the United States has increased significantly since the early 1970s (Karoly, 1993). Most studies show that despite the enviable record of job creation in the United States during the past 20 years (especially in contrast to stagnant job growth in Europe), low-wage jobs (defined variously) account for a high proportion of this employment growth and labor markets show a pronounced polarization of earnings: a growing number of high- and low-wage jobs and a shrinking middle (*Economic Report of the President,* 1994; Freeman & Katz, 1994; Harrison & Bluestone, 1988). For example, the number of year-round full-time workers in the United States earning less than the poverty rate grew by 50% between 1979 and 1992, with almost 20% of all full-time workers now falling into this low-paid category (Dunlop Commission, 1994, p. 18). Average real hourly wages in the United States declined 15% between 1973 and 1993, falling to levels reached in the 1960s. By the 1980s, Canadian researchers also discerned a shift toward greater polarization in the distribution of earnings and stagnation in wages and living standards (although the trends were much less pronounced than in the United States; Economic Council of Canada, 1991; Myles, Picot, & Wannell, 1988).

To what extent has economic globalization promoted this growing labor market polarization?[1] Sassen (1991) argues that globalization promotes dualism in three main ways. First, the leading sectors of global cities—advanced business services and financial instruments that are increasingly traded on world markets—are marked by a highly polarized occupational and wage structure. These sectors contain numerous well-paid bankers, lawyers, and management consultants, large numbers of poorly paid secretaries and routine data-processing workers, but relatively few jobs paying middle-income wages. Second, driven by the "high income lifestyles" of those atop the urban social structure, global cities experience "a process of high-income gentrification which rests on the availability of a vast supply of low-income laborers" (p. 279). Thus, as well-paid "symbolic analysts" (Reich's [1991, p. 282] term) demand low-wage, low-profit services in maintenance, cleaning, and other retail trade, the sectoral transformation of the global city indirectly promotes the creation of low-wage jobs in consumer servicers (custodial, housekeeping, taxi, restaurants, retail sales, parking, etc.), further polarizing the urban social structure. The rapid growth of the low-wage urban tourism and conventions industry in the United States since the 1970s, catering to the travel-leisure proclivities of on-the-move symbolic analysts, is an example of how wage polarization in the leading sectors of urban economies indirectly abets the proliferation of low-wage employment elsewhere.

Global competition—particularly with the development of production facilities in low-wage, less developed countries—promotes a downgrading of urban manufacturing in advanced industrial cities, resulting in downward pressure on wages, the growth of low-cost part-time and contingent labor (part of "flexible production"), and, in some cases, the reemergence of sweatshops (Sassen, 1991; Soja, 1991). In cities such as New York, Los Angeles, and London, relying largely on immigrant laborers, "the decline in wages has reached the point where sweatshop production . . . has become price competitive with cheap imports from Asia" (Sassen, 1991, p. 281). Sweatshops may be an extreme case, but studies show that the distribution of earnings in manufacturing, historically the sector providing substantial middle-income employment opportunities for blue-collar workers in industrial cities, has become increasingly polarized in the United States since the early 1970s (Harrison & Bluestone, 1988).[2]

Globalization may also promote wage polarization by altering the equilibrium between the supply and demand in labor markets (Freeman & Katz, 1994). In an era of modern communications and ever cheaper transportation technologies, production facilities can be built virtually anywhere in the world. As a result, notes Reich (1991), low- and moderate-skill workers are now "in direct competition with millions of routine producers in other nations [who] will happily work for a small fraction of the wages of routine producers in America" (p. 209). This global abundance of labor competing for low- and moderate-skill work exerts downward pressure on wages in advanced societies: Workers in Milwaukee, Detroit, and Baltimore face low-wage competition from Mexico, the Philippines, and Indonesia. Collaterally, globalization widens the market for the "products" of Reich's symbolic analysts; labor markets then translate this increased demand into higher earnings. In short, globalization increases the *supply* of low-skill workers and *demand* for the products of high-skill labor. The result: a bifurcation of earnings and rising inequality.[3]

Global economic integration should not be reified as an uncontrollable force of nature; it involves conscious decisions by corporate executives about the location of investments. Since the early 1970s, the lean and mean business strategies adopted by American corporate managers—redeploying mobile capital, increasingly around the world, in relentless efforts to cut the costs of production—may also have contributed to wage polarization at home (Harrison, 1994). Mollenkopf (1983) persuasively argues that the displacement of investment from the urban Northeast and Midwest toward the Sunbelt in the early 1970s was designed, in part, to exert a "market discipline" on labor and community groups in "liberal" cities (p. 18). Similarly, the globalization of investment—and the credible threat of further globalization—represents a low-wage corporate strategy that undermines the bargaining power of organized labor. Consequently, the capacity of unions to function as vibrant wage-setting institutions, capable of defending wage levels, social programs, and other elements of low- and moderate-skill workers' living standards, has been significantly reduced. Freeman (1993) contends that about 20% of the rise in inequality in the United States since the early 1970s can be attributed to declining union strength, a deunionization that has been facilitated by globalization (p. 134).

There are some important critics of the thesis linking globalization to stagnant and polarizing domestic wages. Krugman (1994), for example, maintains that "the image of the U.S. economy flooded with imports that pay very low wages just isn't right. While the US does some trade with low-wage, Third World countries, the bulk of its trade is with countries that are more or less as advanced as it is and that pay wages close to or even exceeding its own" (p. 147). Although a growing proportion of manufacturing imports come from relatively low-wage developing countries such as China (Dunlop Commission, 1994; Sachs & Shatz, 1994), research by Lawrence and Slaughter (1993) indicates that the typical U.S. trading partner pays workers around 90% of the U.S. wage rate. Lawrence and Slaughter estimate that imports from "low wage" countries, as a percentage of GDP, rose only slightly, from 2.0% in 1960 to 2.7% in 1990 (p. 190). Thus, the 1994 *Economic Report of the President* concluded that "changes in the composition of trade do not appear to explain much of the increase in income inequality" in the United States (p. 119). Rather than focusing on trade and globalization as sources of inequality, economists such as Krugman and Lawrence (1994) emphasize the inequality-inducing effects of new skill-based production technologies (which may provide disproportionate wage benefits to those who master them) and, above all, lagging productivity growth that stunts real wage growth. In the last analysis, however, as Krugman (1994) candidly admits, "we really don't know very well why inequality has increased" (p. 50).

As Howell (1994) has noted, however, focusing narrowly on simply import penetration and export shares misses the key dimension of how globalization engenders wage polarization: by diminishing the bargaining power of low- and moderate-wage labor. Access to global markets and production facilities makes credible corporate threats of plant relocation and demands for worker wage restraint to match low-wage competition, a point conceded in the 1994 *Economic Report of the President.* In the wake of globalization and disinvestment, "concession bargaining" became the norm in numerous industries in the 1970s, so much so that this cost-containment climate spilled over into nontraded sectors and may have helped depress wages in them as well. As one labor economist notes: "The longevity of the [wage] concession movement and its spread to less-than-dire situations suggest that the initial concessions have encouraged other employers to try their luck in demanding similar settlements" (Mitchell, 1985, p. 584). In this fashion, as an integral element in a climate that tips the balance in labor relations against wage earners, globalization may promote polarized wages even in sectors relatively insulated from international trade.

In sum, there are good reasons to believe that globalization has played an important role in spawning wage polarization in U.S. labor markets—even if it is difficult to quantify the significance of such effects. Yet although advanced countries are similarly affected by the growth in international trade—as they are by other potentially polarizing market factors such as shifts in demand for low-skill labor and the introduction of new production technologies—there are distinct differences in labor market outcomes across countries. As Freeman (1994) notes: "The distribution of wages (and incomes) is more unequal in the United States than in other advanced countries. Lower paid workers do much worse compared with the average in the United States than in other advanced countries, while well-paid workers do much better" (p. 12). If globalization and other market factors are integral in producing wage polarization, roughly similar wage structure outcomes should be observed in advanced countries. Thus, these apparent differences draw attention to wage-setting systems: To what extent do different labor market rules and social institutions affect the wage polarization tendencies that apparently accompany globalization?

Wage Polarization in Selected U.S. and Canadian Cities: 1970 to 1990

Researchers have used a variety of measures to gauge changes in wage inequality or polarization through time. One technique is to measure wage dispersion: for example, to examine the gap in average earnings between the lowest and highest deciles in a given wage distribution (Freeman & Katz, 1994; Goldin & Margo, 1992; Karoly, 1993). If, for example, the gap increases between the average wages paid to the top 10% of wage earners and those paid to the bottom 10%, researchers generally conclude that inequality is increasing.

Although the wage dispersion approach is a perfectly fine measure of inequality, it may not capture precisely what is of real interest: changes in the actual well-being of workers. For example, in an economy in which overall real earnings are rising rapidly, the real wages—and hence the living standards—of low-paid workers and middle-income workers may be improving considerably, even if the income *differentials* between the highest and lowest paid workers are also increasing.[4] Conversely, even if wage dispersion is shrinking in a particular society, if real wages are declining, the overall well-being of most workers is probably declining. Thus, to examine the influence of globalization on actual living standards, the dispersion measure is probably not the best gauge of wage polarization.

Similarly problematic is a second popular measure of changing wage distributions. This technique divides wage earners into earnings classes, defined as a specified percentage of median earnings. For example, the 1991 Economic Council of Canada study, *Employment in the Service Economy,* breaks wage earners into three classes: a middle-class category (all workers earning 75% to 125% of median earnings), a lower category earning below 75% of the median, and an upper category earning more than 125% of the median. Again, this is a standard measure of an earnings distribution, but it can badly distort actual trends in living standards when real earnings are declining or rising. For example, if the real median wage has declined over a period, a finding that the proportion of workers falling into the middle-earnings category (75% to 125% of the median, in the example above) is unchanged during that same period will accurately convey a stable distribution of wages but mask a deterioration in the living standards of workers.

To avoid these methodological pitfalls, a better way to examine the polarization of wages is to establish, in real dollars, earnings categories that reflect actual standards of living and then to examine through time whether the number of workers in these categories has expanded or contracted.[5] As the following six tables indicate, I have divided workers in selected U.S. and Canadian cities into three earnings classes: a low-wage category (workers earning under $20,000 annually in 1990 constant dollars), a middle-class category (workers earning between $20,000 and $39,999), and an upper-earnings strata (workers earning over $40,000 a year). Although all wage-earner classification schemes have a certain arbitrariness, this one makes sense. As Schwarz and Volgy (1993) have persuasively argued, $20,000 annually (in 1990 dollars) is the appropriate cutoff for a low-economy, no-frills family budget: For an urban household of four, this includes minimal food, housing, clothing, transportation, and medical expenditures, with nothing left over for such frills as entertainment or vacations. The middle category roughly reflects earnings equal to about 33% above and below the median urban family expenditures in the United States in 1990; the upper-earnings categories includes all other workers.

To sidestep any distortions that might result from central city-suburban disparities, earnings data have been collected for metropolitan areas. The metropolitan areas examined include Canada's three largest urban centers and a geographically varied sample of cities from across the United States, reflecting fast-growing Sunbelt metropolises, older, industrial centers of the Northeast and Midwest, and cities widely thought to be at the cutting edge of global restructuring.

Tables 4.1 through 4.6 array data from these metropolitan areas on the distribution of workers, by wage-earning classes, between 1970 and 1990. Tables 4.2 and 4.5 show what proportion of net employment growth was accounted for by each wage level between 1970 and 1990, an indicator of the quality of jobs that have been generated in U.S. and Canadian cities since 1970.[6]

The U.S. metropolitan areas uniformly experienced a surge in low-wage employment during the 1970s. As Table 4.1 reveals, in deindustrializing cities in the Northeast and Midwest, the low-wage share of total employment grew substantially during the 1970s. Chicago was typical, as the low-wage proportion of total employment jumped from 38.9% in 1970 to 43.6% in 1980. In Baltimore, Boston, Chicago, Detroit, and Milwaukee, over two thirds of the net employment growth between 1970 and 1980

TABLE 4.1 Low Wages in U.S. Cities, 1970-1990: Percentage of Total Workforce Earning Under $20,000 Annually (in 1990 Constant Dollars): Selected Metropolitan Areas

Metro Area	*1970*	*1980*	*1990*
Atlanta	46.0	49.2	45.8
Baltimore	45.6	48.4	46.3
Boston	43.0	48.3	39.5
Charlotte	52.8	54.6	51.7
Chicago	38.9	43.6	44.7
Dallas	47.6	49.1	49.1
Detroit	34.6	38.8	44.6
Los Angeles	39.8	46.1	46.0
Miami	53.5	53.1	56.0
Milwaukee	41.9	45.6	50.8
Minneapolis	43.5	47.4	46.9
Seattle	36.9	42.5	44.4

SOURCE: U.S. Bureau of the Census, 1973, 1984; U.S. Department of Transportation, Bureau of Transportation Statistics, 1994.

came in jobs paying under $20,000 annually (in 1990 constant dollars), a veritable proliferation of low-wage employment.

These figures, though striking, are hardly surprising: The loss of middle-class manufacturing jobs, combined with overall urban economic stagnation in the Northeast and Midwest, has clearly been associated with declining living standards since the early 1970s (Bluestone & Harrison, 1982; Stanback & Noyelle, 1982). These cities, some analysts might argue, became uncompetitive in the increasingly globalized economy of the 1970s, and their workers paid the price in shrinking real wages and job opportunities. Not so coincidentally, between 1970 and 1980 exports and imports as a proportion of GDP in the United States surged from 11.2% to 21.1% (*Economic Report of the President,* 1994, p. 269). In addition, as highly unionized and politically "contested" terrain, to use Mollenkopf's (1983) description, these cities were also the prime sites for mobile capital to use globalization to assert "market discipline" over urban "liberalism" (p. 18).

U.S. cities generally considered globally competitive, however, also experienced a surge in low-wage job creation during the 1970s. Miami is the chief U.S. urban link to Latin America and the Caribbean, whereas Seattle and Los Angeles are two West Coast cities with growing Pacific Rim linkages; all would be considered more successfully globally integrated than, say, Detroit or Milwaukee. Yet as Table 4.2 shows, well over

TABLE 4.2 Wage Polarization in U.S. Cities Since 1970: Percentage of Net Employment Growth Accounted for by Earnings Strata (in 1990 Constant Dollars): Selected Metropolitan Areas

Metro Area	*Earnings Strata*	*1970-1980*	*1980-1990*	*1970-1990*
Atlanta	$0-$19,999	55.2	40.6	45.6
	$20-$39,999	36.8	36.2	36.4
	Over $40,000	8.0	23.4	18.0
Baltimore	$0-$19,999	72.5	39.9	47.7
	$20-$39,999	26.8	31.8	30.6
	Over $40,000	0.7	28.4	21.8
Boston	$0-$19,999	78.3	24.5	35.5
	$20-$39,999	30.0	42.4	39.8
	Over $40,000	*	33.1	24.7
Charlotte	$0-$19,999	57.1	46.3	51.1
	$20-$39,999	36.5	33.3	34.8
	Over $40,000	6.4	20.1	14.0
Chicago	$0-$19,999	76.2	49.3	58.8
	$20-$39,999	23.4	12.9	16.6
	Over $40,000	0.4	28.2	18.4
Dallas	$0-$19,999	50.5	49.0	49.7
	$20-$39,999	40.5	29.4	35.1
	Over $40,000	9.1	21.6	15.2
Detroit	$0-$19,999	67.2	76.4	72.7
	$20-$39,999	21.0	0.9	9.1
	Over $40,000	11.8	22.7	18.2
Los Angeles	$0-$19,999	50.5	49.0	49.7
	$20-$39,999	40.5	29.4	35.1
	Over $40,000	9.1	21.6	15.2
Miami	$0-$19,999	52.3	63.0	58.5
	$20-$39,999	39.2	19.3	27.7
	Over $40,000	8.6	17.6	13.8
Milwaukee	$0-$19,999	69.6	83.2	77.4
	$20-$39,999	34.4	*	13.5
	Over $40,000	*	20.5	9.2
Minneapolis	$0-$19,999	57.6	45.2	51.0
	$20-$39,999	36.6	30.7	33.5
	Over $40,000	5.8	24.1	15.6
Seattle	$0-$19,999	55.7	48.2	51.1
	$20-$39,999	35.3	31.7	33.1
	Over $40,000	10.1	20.1	16.2

SOURCE: U.S. Bureau of the Census, 1973, 1984; U.S. Department of Transportation, Bureau of Transportation Statistics, 1994.
*Strata declined during this period.

half the net jobs created in these cities during the 1970s fell into the low-wage category: In Los Angeles, the low-wage share of total employment grew from 39.8% to 46.1% during the decade, whereas in Seattle the low-wage component increased from 36.9% to 42.5%. These data suggest that "adapting to global realities" was hardly an economic development strategy that avoided the snare of low-paying jobs or paid big dividends for the majority of local workers.

Although employment grew much more rapidly in U.S. metropolitan areas in the 1980s than in the 1970s, wage polarization generally solidified during the last decade. In most cities, more than 40% of the net jobs created during the 1980s were low wage, with over three quarters of the net job growth falling into this category in strongly deindustrializing Milwaukee and Detroit. Although the rate of low-wage job creation slowed somewhat in most cities during the 1980s, the rate of globalization of the economy also decelerated during the decade: The ratio of trade to GDP virtually doubled during the 1970s, whereas it increased by only 0.3% between 1980 and 1990.

Unlike the 1970s, during the 1980s there was also strong growth in high-wage job creation in most metropolitan areas, particularly in redeveloping downtown business services centers. In Baltimore, to take a typical example, although jobs paying over $40,000 annually (in 1990 dollars) accounted for only 0.7% of net job growth during the 1970s, they represented 28.4% of net employment created during the 1980s. Thus, as Table 4.3 clearly shows, with growth at the top and bottom of the wage distribution, there was a continuing shrinkage in the proportion of middle-class wage earners through the 1980s and a consolidation of labor market polarization.

As was the case during the 1970s, wage polarization during the 1980s was not limited simply to declining or deindustrializing cities. The greatest share of net employment growth in the 1980s in globally competitive cities such as Seattle and Los Angeles was in low-wage jobs. Indeed, almost two thirds of the net employment growth in Miami was in the low-wage category, a rate exceeded only by Milwaukee and Detroit in the sample of cities examined here.[7]

The one clear exception to the U.S. pattern was metropolitan Boston where, although low-wage jobs proliferated during the 1970s, the low-wage share of total employment shrank precipitously from 48.3% to 39.5% between 1980 and 1990. With its concentration of world-class research universities, Boston was uniquely positioned to take advantage of a confluence of trends during the 1980s: surging defense-related research and development spending, high-technology production and engineering, and the concomitant expansion of knowledge-intensive industries (Rosegrant &

TABLE 4.3 A Shrinking Middle? Percentage of Workforce Earning Between $20,000 and $39,999 Annually (in 1990 Constant Dollars)

Metro Area	*1970*	*1980*	*1990*
Atlanta	39.8	38.7	37.7
Baltimore	41.4	39.9	37.8
Boston	41.0	39.3	40.5
Charlotte	36.3	36.4	35.3
Chicago	43.3	40.7	37.3
Dallas	36.3	38.5	35.5
Detroit	43.8	40.9	34.7
Los Angeles	41.8	38.6	34.0
Miami	34.0	35.5	30.9
Milwaukee	43.7	42.5	36.1
Minneapolis	40.9	39.7	37.5
Seattle	44.8	42.0	38.7

SOURCE: U.S. Bureau of the Census, 1973, 1984; U.S. Department of Transportation, Bureau of Transportation Statistics, 1994.

TABLE 4.4 Low Wages in Canadian Cities, 1970-1990: Percentage of Total Workforce Earning Under $20,000 Annually (in 1990 Constant Canadian Dollars)

Metro Area	*1970*	*1980*	*1985*	*1990*
Montreal	53.8	44.8	48.6	46.1
Toronto	50.5	42.9	46.2	39.0
Vancouver	51.4	44.0	48.3	45.3

SOURCE: Statistics Canada, 1975, 1983a, 1983b, 1983c, 1993b.

Lampe, 1992). As a result, Boston's superheated economy achieved virtual full employment during the 1980s, creating a tight labor market that drove up wages across-the-board (Osterman, 1991). For example, during the mid-1980s heyday of the "Massachusetts Miracle," labor shortages in the Boston area led McDonald's to offer starting wages of $7.00 an hour to entice workers—well above the statutory minimum wage (Krugman, 1994, p. 209). Higher entry-level wages apparently help push up earnings throughout the wage structure.

As Tables 4.4, 4.5, and 4.6 indicate, wage trends in the urban labor markets of Canada differed considerably from the U.S. pattern (excluding 1980s Boston). In all three major Canadian urban centers, the low-wage component of the labor market shrank considerably during the 1970s. Only

TABLE 4.5 Wage Polarization in Canadian Cities Since 1970: Percentage of Net Employment Growth Accounted for by Earnings Strata (in 1990 Constant Canadian Dollars)

Metro Area	*Earnings Strata*	*1970-1980*	*1980-1985*	*1985-1990*	*1970-1990*
Montreal	$0-$19,999	*	89.8	26.5	28.8
	$20-$39,000	69.1	*	41.3	34.5
	Over $40,000	59.7	30.6	32.1	36.7
Toronto	$0-$19,999	28.4	65.2	*	27.8
	$20-$39,999	40.8	10.7	53.2	37.3
	Over $40,000	30.8	24.0	61.6	34.8
Vancouver	$0-$19,999	32.3	70.2	32.8	40.3
	$20-$39,000	38.1	*	40.1	30.6
	Over $40,000	29.6	31.7	27.2	29.2

SOURCE: Statistics Canada, 1975, 1983a, 1983b, 1983c, 1993b.
*Strata declined during this period.

TABLE 4.6 A Tale of Four Cities: Wage Distribution Trends in Chicago, Montreal, Seattle, and Toronto Since 1970: Percentage of All Workers in Wage-Earning Classes

Metro Area	*Earnings Strata*	*1970*	*1980*	*1990*
Chicago	$0-$19,999	38.9	43.6	44.7
	$20-$39,000	43.3	40.7	37.3
	Over $40,000	17.8	15.7	18.0
Montreal	$0-$19,999	53.8	44.8	46.1
	$20-$39,000	36.2	39.8	35.9
	Over $40,000	10.0	15.4	18.0
Seattle	$0-$19,999	36.9	42.5	44.4
	$20-$39,999	44.8	42.0	38.7
	Over $40,000	18.3	15.5	16.9
Toronto	$0-$19,999	50.5	42.9	39.0
	$20-$39,999	38.3	40.4	37.8
	Over $40,000	11.0	16.7	23.2

SOURCE: Statistics Canada, 1975, 1983a, 1983b, 1983c, 1993b; U.S. Bureau of the Census, 1973, 1984; U.S. Department of Transportation, Bureau of Transportation Statistics, 1994.

about a quarter of the net employment growth in Toronto and Vancouver during the decade was in low-wage jobs, whereas Montreal experienced a net loss of low-wage employment during the 1970s. All three cities experienced moderate growth in the middle-wage category and substantial gains in the high-wage group. This depolarization of the Canadian urban wage structure occurred during a decade in which trade as a proportion of GDP increased from 42.5% to 54.7%, suggesting that certain Canadian policies and institutions may have substantially mitigated the polarizing consequences of globalization on urban labor markets (Statistics Canada, 1991, p. 5). Clearly, if globalization were an all-determining factor shaping urban wage structures, in view of the far greater importance of international trade in the Canadian economy compared with the U.S. economy, greater polarization in Canadian cities might have been anticipated; quite the opposite was the case, however.

During the first half of the 1980s, however, wage trends in Canadian cities began resembling those in the United States. This was not because of a surge in low-wage international trade: The ratio of trade to Canadian GDP actually shrank slightly between 1980 and 1985 and by almost 9% between 1985 and 1990. Clearly, the deep recession of 1981 to 1983 severely eroded real wages across all earnings classes in urban Canada. In addition, as I examine shortly, policy shifts during the 1980s may also have contributed to growing wage polarization: chiefly, cuts in social programs, attacks on unions, and a decrease in the real value of the minimum wage in several provinces. Whatever the reasons, by the mid-1980s, there appeared to be a certain Americanization of the wage structure in metropolitan Canada. As Table 4.5 reveals, however, by end of decade the Canadian wage structure appeared to return to historical patterns, with a rate of low-wage job creation in Toronto and Montreal again below the U.S. rate—perhaps a reflection of the underlying "worker friendly" labor market institutions in Canada that withstood some of the policy shifts of the 1980s.[8]

Table 4.6 highlights the differences in wage trends in U.S. and Canadian cities since 1970. In both the older, deindustrializing city (Chicago), and the city whose growth is tied to increasing international trade (Seattle), the American urban wage structure became considerably more polarized between 1970 and 1990. By contrast, both Montreal and Toronto witnessed a significant depolarization of wages during the 1970s, a trend that continued in Toronto during its extraordinary growth following the early 1980s recession.

Differences in U.S.-Canadian Wage Polarization: Some Explanations

How can one account for these differences in the labor market outcomes in Canadian and U.S. cities? Recent research suggests that small differences in institutions and public policy matter in explaining variations in wage structures (Card & Freeman, 1994). The more generous "social wage" in Canadian provinces, consisting principally of more liberal income-maintenance, family-allowance, and unemployment compensation programs, has been cited as muting tendencies toward wage polarization and inequality (Bakker, 1991).[9] Since the early 1970s, the minimum wage in Ontario, Quebec, and British Columbia has been higher, relative to average wages, than in the United States, increasing the wage floor in Canadian cities (Government of Canada, Ministry of Labor, 1993). Finally, union density remains much higher in Canada than in the United States. In Canada, union membership in 1992 was 29.7%, only a slight fall from its 1978 peak of 31.2%. By contrast, the United States experienced precipitous deunionization, with union membership declining from 27% in 1970 to 15.8% in 1992. Indeed, Lemieux (1994) contends that 40% of the difference in the growth of income inequality between the United States and Canada during the 1980s was attributable to variations in levels of unionization (p. 97).

The interplay of these factors in influencing wage trends can be most clearly observed in the case of Montreal. By most structural indicators, Montreal should have exhibited wage trends similar to U.S. metropolitan areas. Between 1971 and 1991, the manufacturing share of total employment in metropolitan Montreal declined from 30.7% to 18.8%. Unemployment rose from 6.5% in 1975 to 9.0% in 1980 and to 12.4% in 1990. Indeed, an impressive and extensive body of research now documents the "decline" of Montreal in the wake of economic restructuring since the late 1960s (Coffey & Polèse, 1993; Côté, 1990; Lamonde & Martineau, 1992; Polèse, 1990).

Yet despite these economic problems, metropolitan Montreal's labor market manifested a discernible diminution in low-wage employment and broadening of the middle during the 1970s. Moreover, after the disastrous, wage-polarizing recession of 1982 to 1983, a moderate wage depolarization had reemerged by the late 1980s.

Several policies and labor market institutions appear responsible, at least in part, for these trends. During the 1970s, with the expansion of the provincial and municipal governments as well as a surge in public works investments associated with transportation (the *Métro*) and the 1976 Olympics, Montreal witnessed a significant increase in well-paid public employment—the type of nonmarket services that historically led the way, for example, in Sweden's egalitarian wage distribution (Esping-Anderson, 1990). In addition, through the early 1980s, Quebec had the highest minimum wage in Canada. During the 1970s, although the U.S. minimum wage lost 37% of its real value, the inflation-adjusted minimum wage in Quebec increased by 14%. With well-paid public employment "pulling" wages and a generous minimum wage "pushing" them upward, the low-wage component of the Montreal labor market shrank, in relative terms, whereas the middle and top expanded.

Last and most important, Montreal retained a high rate of unionization (35.3% in Quebec in 1980, compared with 20.1% in the United States). Quebec labor was strengthened during the 1970s by legislation such as the 1977 antistrikebreaker law. Moreover, Quebec's powerful and expanding public employees unions pursued an explicitly *égalitariste* wage policy during the 1970s, a strategy designed to reduce wage gaps and promote greater equity in the labor market. According to one study, the result of this strategy was a reduction of 25% between 1971 and 1983 in the gaps separating the highest- and lowest-paid workers in the public employment salary structure (Beaucage, 1989, p. 67).

It is difficult to disentangle the impact of these policies and institutions from other market influences in shaping Montreal's wage structure. Increases in overall levels of education and adjustments in the city's linguistic hierarchy during the 1970s undoubtedly played an important role in depolarizing the city's labor market. In addition, unlike many U.S. cities, Montreal's manufacturing was historically composed of low-wage, low-productivity industries, such as textiles, leather, clothing, food, and beverages. Thus, although deindustrialization was stripping many U.S. cities of well-paid blue-collar jobs in steel and automobile production, in many cases industrial restructuring in Montreal has meant shedding low-wage jobs and replacing them with better paying service-sector employment that, overall, has had a depolarizing impact on the city's wage distribution. For example, in 1990, 41.3% of Montreal's manufacturing workers earned under $20,000 a year, compared with 34.7% of workers in nonmarket services (Statistics Canada, 1993a).

By the mid-1980s, as I have noted, the strong depolarization of Montreal's wage structure ended. Persistent double-digit unemployment created a slack labor market, undercutting workers' bargaining power and eroding real wages. Employment growth and wage increases in the public sector, a driving force in the "egalitarianization" of wages in the 1970s, slowed considerably. There was a general curtailment of Quebec's social wage during the decade. Moreover, although union membership remained high, at 38% of the labor force in 1987, Quebec unions were put on the defensive during the decade by both the provincial government and more aggressive private employers. In short, during the 1980s there was something of an

Americanization of Quebec and Canadian labor relations and public policy. As one example, the real value of the minimum wage plummeted 22% in Quebec during the 1980s, comparable to the 29% decline experienced in the United States. All of these trends, combined with an ongoing deindustrialization that began to hit some of Montreal's higher value-added industrial sectors, took a toll on the city's wage structure.

The degree to which different labor market rules are responsible for varying wage structures is difficult to determine with quantitative precision. There are simply too many possible causes of variations in labor market outcomes. Nevertheless, the Montreal case, combined with general differences in labor market rules between the United States and Canada, suggests that policies and institutions are important in shaping labor market outcomes. Although similar globalization pressures affected cities on both sides of the border, during the 1970s there were important variations in labor market rules—and demonstrably different wage distribution outcomes. During the 1980s, as some of these policy differences attenuated, so too did some of the variations in wage structure between cities in the two countries, although underlying differences between the two labor market systems left Canada's wage structure markedly less polarized.

The U.S.-Canadian comparison underscores the need to allow sufficient explanatory room for public policies, social organizations, and institutional arrangements in the analysis of urban labor market outcomes. As Myles (1991) has argued, there is nothing intrinsic about economic restructuring "dictating that it will be dominated by low wage, low skill jobs and intermittent employment" (p. 356). The Canadian-U.S. comparison hints that the generosity of social wages, levels of unionization, and whether well-paid public employment is a leading economic sector are all among the institutional factors mediating the labor market impact of globalization.

The value and, hence, remuneration attached to various jobs is not exclusively market driven. As Molotch (1991) notes, "The income or dignity of a job has never been intrinsic to the task performed, but a matter of institutional definition" (p. 177). The "good" manufacturing jobs disappearing from U.S. cities under the pressures of globalization did not pay well because they were intrinsically "high value": They became good jobs because unionized workers mobilized and were able to secure a higher proportion of industry revenues. Similarly, as Myles contends, service-sector wages are not purely market driven. Employment in the day-care industry, for example, can be structured as low-wage, unskilled baby-sitting services as in the United States and Canada, or as high-wage, skilled educational services as in France and Sweden. That is a matter of social decision and public policy, not an immutable law of value that says child-care services are "bad" jobs. Globalization and other economic factors do create a general environment in which urban labor markets must function, but policymakers appear to possess more market-shaping maneuverability than is commonly believed (Logan & Swanstrom, 1991).

■ Conclusion

Trends in North American urban wage structures suggest that adapting to global realities is not a surefire route to urban prosperity. *How* cities and businesses adjust to globalization—whether, for example, competitiveness is based on lowering standards of living to compete with low-wage workers overseas—matters in determining how globalization will restructure urban economies (Drache, 1992; Drache & Gertler, 1991; Mackenzie, 1992). As Claude Béland, the president of Quebec's powerful financial institution, the *Mouvement Desjardins,* has astutely pointed out, a fetish about global competitiveness may overshadow the most important policies affecting local economic well-being: neighborhood and regional development strategies (Shalom, 1993).

On the other hand, a more sober globalization scenario envisions an international "race to the bottom" in which advanced societies are virtually powerless to protect living standards in the face of fiercely competitive international pressures. As Simeon (1991) puts it, there is a "growing mismatch between the scope and scale of the issues with which we must cope and the reach of political institutions through which we must deal with them" (p. 48). These pressures are undeniably real. Yet comparisons of U.S. and Canadian labor markets, as well as U.S.-European comparisons by Freeman and associates, reveal that wage polarization is not an inevitable by-product of globalization; much depends on how the effects of globalization are filtered through a society's labor market system.

Unquestionably, the great conundrum for labor market policy in the globalized 1990s is how to balance (a) sufficient market flexibility to create jobs and (b) social institutions to protect against polarization and inequality. Concern over "Eurosclerosis"—high unemployment rates in Europe allegedly flowing from high labor costs and market rigidities—has led European Community leaders to endorse proposals to lower minimum wages and cut nonwage labor costs that finance comprehensive social welfare programs (Organization for Economic Cooperation and Development [OECD], 1994). Social policy reform is also on Canada's political agenda. Making job creation cheaper in a competitive international trading system is at the heart of these proposals.

The jobs-wages trade-off cannot be ignored. At least some of the difference in the unemployment rates in the United States, Europe, and Canada, however, reflects whether individuals not working are likely to be classified as unemployed (more in Europe and Canada, less in the United States) rather than massive variations in unemployment. Moreover, comparative labor market research suggests that the costs in labor market efficiency of European and Canadian-style social protection are often exaggerated (Blank, 1994).

On the other hand, as the data on U.S.-Canadian urban wage structures would suggest, trimming the social wage and moving toward a U.S.-style labor market system would entail significant costs in increased wage polarization and social inequality. Solving the OECD employment crisis by generating a slew of low-wage, part-time jobs hardly makes sense. Moreover, overemphasizing low-cost exports at the expense of a broad, consumer-driven domestic market introduces greater cyclical instability to domestic economies and will further fracture the middle class and exacerbate social polarization. In the last analysis, absent a "Global New Deal" or "social charter" that sets employment standards for the international trading system, urban and national labor market systems with high rates of unionization and generous social wages appear to offer the best safeguards against the polarization tendencies of the global economy.[10]

NOTES

1. Globalization is one among many potential causes of wage polarization identified by recent research. Others include changes in labor supply (changing rates of growth among college- and non-college-educated labor); demographic shifts (the impact of baby boomers and, particularly, more women entering the labor force in the 1970s); new production technologies (such as computers, which may give wage premiums to skilled workers); and immigration (which increases the supply of low-skilled labor and hence lowers wages for entry-level workers). For an excellent overview and effort to weight these causes, see Freeman and Katz, 1994.

2. In other words, wage polarization cannot be attributed simply to sectoral shifts—to the disappearance of good manufacturing jobs and the rise of low-paid services. *Within*-sector wage polarization has also increased.

3. Virtually all studies have noted the huge earnings gap opening in the late 1970s in the United States between college- and non-college-educated labor. Part of the explanation for this gap, as Freeman and Katz (1994) note, was a relative deceleration in the growth of college graduates starting in the 1970s, thus decreasing the relative supply of this group and raising its "price" (wages). As noted earlier, there may have been a "technology premium"

earned by college-educated workers as well, although the evidence is not conclusive. The influence of globalization on labor supply and demand, however, exacerbated these internally generated differentials. In addition, recent research reveals a decline in wages among the college educated beginning in the late 1980s, suggesting that globalization and other competitive factors may now be adversely affecting even the well educated.

4. This is precisely what occurred, for example, in Japan during the 1980s.

5. An additional conceptual methodological problem should be noted. Most analysts agree that when researching wage polarization, it is best to collect data on full-time, full-year workers to control for voluntarily part-time workers whose lower wages may distort downward the overall distribution. Unfortunately, the urban wages data for the United States are available only for all workers and are not disaggregated by full- and part-time workers. I have examined the differences between the wage distribution in Canadian cities for all workers and for only full-time, full-year workers, however; the trends for both groups are identical. Thus, the data problem may not be as serious as it appears.

6. Net employment change is calculated as follows:

$$\text{Employment in: } \frac{\text{stratum } i \text{ in year } t_2 - \text{stratum } i \text{ in year } t_1}{\text{all strata in year } t_2 - \text{all strata in in year } t_1}$$

(See Harrison & Bluestone, 1988, p. 198)

7. At the national level in the United States, some studies calculate that roughly one fifth of the growth in the wage gap between skilled and unskilled workers since the 1970s can be attributed to the growing supply of unskilled immigrant labor. See, for example, Borjas, Freeman, and Katz (1992). Such effects may be even more discernible in highly concentrated immigration settings such as Miami and Los Angeles, in which the willingness of immigrants to work in low-wage, sweatshop conditions lowers the floor for overall wages. More detailed local labor market studies will be necessary to confirm this speculation.

8. These wage trends have undoubtedly taken a turn for the worse in the early 1990s, as Toronto and Montreal have seen their employment bases devastated by a deep and lingering "restructuring" recession.

9. The social wage measure calculates the net impact of government spending and taxes on workers' standard of living. In general terms, it is calculated by subtracting taxes paid by workers from the wages and benefits received from the state (with similar calculations done for capital). For a description of the methodology, as well as a brief comparison of the U.S. and Canadian social wage, see Bakker (1991).

10. For an example of efforts to forge such a "social charter" across Canada, see Echenberg et al. (1992).

REFERENCES

Bakker, I. (1991). Canada's social wage in an open economy, 1970-1983. In D. Drache & M. Gertler (Eds.), *The new era of global competition: State policy and market power* (pp. 270-287). Montreal, PQ: McGill-Queen's University Press.

Beaucage, A. (1989). *Syndicats, salaires, et conjuncture économique: L'expérience des fronts communs du secteur public québécois de 1971 à 1983* [Unions, wages, and the state of the economy: The experience of the public sector common front in Quebec from 1971 to 1983]. Quebec: Les Presses de l'Université du Québec.

Blank, R. (1994). *Social protection vs. economic flexibility: Is there a tradeoff?* Chicago: University of Chicago Press.

Bluestone, B., & Harrison, B. (1982). *The deindustrialization of America.* New York: Basic Books.

Borjas, G. J., Freeman, R. B., & Katz, L. F. (1992). On the labor market effects of immigration and trade. In G. Borjas & R. B. Freeman (Eds.), *Immigration and the work force* (pp. 213-244). Chicago: University of Chicago Press.

Card, D., & Freeman, R. B. (1994). Introduction. In D. Card & R. B. Freeman (Eds.), *Small differences that matter: Labor markets and income maintenance in Canada and the United States* (pp. 1-19). Chicago: University of Chicago Press.

Coffey, W., & Polèse, M. (1993). Le déclin de l'empire montréalaise: Regard sur l'économie d'une metropole en mutation [The decline of the Montreal empire: A look at a metropolitan economy in transformation]. *Recherches sociographiques, 34,* 417-438.

Commission on the Future of Worker-Management Relations [Dunlop Commission]. (1994). *Fact finding report.* Washington, DC: U.S. Department of Labor and U.S. Department of Commerce.

Côté, M. (1990). *Un cadre d'analyse pour le Comité ministériel permanent de développement du Grand Montréal* [An analytic framework for the permanent ministerial committee on the development of greater Montreal]. Montreal, PQ: SECOR.

Drache, D. (1992). Introduction. In D. Drache (Ed.), *Getting on track: Social democratic strategies for Ontario* (pp. xiii-xviii). Montreal, PQ: McGill-Queen's University Press.

Drache, D. & Gertler, M. (1991). The world economy and the nation-state: The new international order. In D. Drache & M. Gertler (Eds.), *The new era of global competition: State policy and market power* (pp. 3-25). Montreal, PQ: McGill-Queen's University Press.

Echenberg, H., Milner, A., Myles, J., Osberg, L., Phipps, S., Richards, J., & Robson, W. B. P. (1992). *A new social charter for Canada? Perspectives on the constitutional entrenchment of social rights.* Toronto, ON: C. D. Howe Institute.

Economic Council of Canada. (1991). *Employment in the service economy.* Ottawa, ON: Minister of Supply and Services.

Economic Report of the President. (1994). Washington, DC: Government Printing Office.

Esping-Anderson, G. (1990). *Three worlds of welfare capitalism.* Princeton, NJ: Princeton University Press.

Freeman, R. B. (1993). How much has de-unionization contributed to the rise in male earnings inequality? In S. Danziger & P. Gottschalk (Eds.), *Uneven tides: Rising inequality in America* (pp. 133-164). New York: Russell Sage.

Freeman, R. B. (1994). How labor fares in advanced economies. In R. B. Freeman (Ed.), *Working under different rules* (pp. 1-28). New York: Russell Sage.

Freeman, R. B., & Katz, L. F. (1994). Rising wage inequality: The United States and other advanced countries. In R. B. Freeman (Ed.), *Working under different rules* (pp. 29-62). New York: Russell Sage.

Goldin, C., & Margo, R. (1992). The great compression: The wage structure in the United States at mid-century. *Quarterly Journal of Economics, 107,* 1-34.

Government of Canada, Ministry of Labor. (1993). *Employment standards legislation in Canada.* Ottawa, ON: Minister of Supply and Services.

Gudell, H. (1994). Metro regional strategies must take global view. *Economic Developments, 19*(2), 6-7.

Harrison, B. (1994). *Lean and mean: The changing landscape of corporate power in the age of flexibility.* New York: Basic Books.

Harrison, B., & Bluestone, B. (1988). *The great U-turn: Corporate restructuring and the polarizing of America.* New York: Basic Books.

Howell, D. R. (1994). The skills myth. *The American Prospect, 18,* 81-90.

Karoly, L. A. (1993). The trend in inequality among families, individuals, and workers in the United States: A twenty five year perspective. In S. Danziger & P. Gottschalk (Eds.), *Uneven tides: Rising inequality in America* (pp. 19-98). New York: Russell Sage.

Krugman, P. (1994). *Peddling prosperity: Economic sense and nonsense in the age of diminished expectations.* New York: Norton.

Krugman, P. R., & Lawrence, R. Z. (1994, April). Trade, jobs, and wages. *Scientific American, 270,* 44-49.

Lamonde, P., & Martineau, Y. (1992). *Désindustrialisation et restructuration économique: Montréal et les autres grandes métropoles nord-américaines, 1971-1991* [Deindustrialization and economic restructuring: Montreal and other large North American metropolises]. Montreal, PQ: INRS-Urbanisation.

Lawrence, R. Z., & Slaughter, M. J. (1993). International trade and American wages in the 1980s: Giant sucking sound or small hiccup. *Brookings Papers on Economic Activity, 2,* 161-210.

Lemieux, T. (1994). Unions and wage inequality in Canada and the United States. In D. Card & R. B. Freeman (Eds.), *Small differences that matter: Labor markets and income maintenance in Canada and the United States* (pp. 69-108). Chicago: University of Chicago Press.

Logan, J. R., & Swanstrom, T. (1991). Urban restructuring: A critical view. In J. R. Logan & T. Swanstrom (Eds.), *Beyond the city limits* (pp. 3-24). Philadelphia: Temple University Press.

Mackenzie, H. (1992). Dealing with the new global economy: What the premier's council overlooked. In D. Drache (Ed.), *Getting on track: Social democratic strategies for Ontario* (pp. 5-16). Montreal, PQ: McGill-Queen's University Press.

Mitchell, D. J. B. (1985). Shifting norms in wage determination. *Brookings Papers on Economic Activity, 2,* 575-599.

Mollenkopf, J. (1983). *The contested city.* Princeton, NJ: Princeton University Press.

Molotch, H. (1991). Urban deals in comparative perspective. In J. R. Logan & T. Swanstrom (Eds.), *Beyond the city limits* (pp. 175-198). Philadelphia: Temple University Press.

Myles, J. (1991). Post-industrialism and the service economy. In D. Drache & M. Gertler (Eds.), *The new era of global competition: State policy and market power* (pp. 351-366). Montreal, PQ: McGill-Queen's University Press.

Myles, J., Picot, G., & Wannell, T. (1988). *Wages and jobs in the 1980s: Changing youth wages and the declining middle.* Ottawa, ON: Statistics Canada.

Organization for Economic Cooperation and Development. (1994). *The OECD jobs study: Facts, analysis, strategies.* Paris: Author.

Osterman, P. (1991). Gains from growth: The impact of full-employment on poverty in Boston. In C. Jencks & P. E. Peterson (Eds.), *The urban underclass* (pp. 122-134). Washington, DC: Brookings Institution.

Peirce, N. R. (1993). *Citistates: How urban America can prosper in a competitive world.* Washington, DC: Seven Locks Press.

Polèse, M. (1990). La thèse du déclin économique de Montréal: Revue et corrigée [The thesis of the economic decline of Montreal: Reviewed and corrected]. *L'Actualité économique, 66*(2), 133-146.

Reich, R. (1991). *The work of nations.* New York: Knopf.

Rosegrant, S., & Lampe, D. (1992). *Route 128: Lessons from Boston's high tech community.* New York: Basic Books.

Sachs, J. D., & Shatz, H. J. (1994). Trade and jobs in U.S. manufacturing. *Brookings Papers on Economic Activity, 1,* 1-69.

Sassen, S. (1991). *Global city: New York, London, and Tokyo.* Princeton, NJ: Princeton University Press.

Schwarz, J. E., & Volgy, T. (1993). *The forgotten Americans: Thirty million working poor in the land of opportunity.* New York: Norton.

Shalom, F. (1993, November 30). Quebec held back by two-tiered vision, Caisse chief says. *The Montreal Gazette,* p. C3.

Simeon, R. (1991). Globalization and the Canadian nation-state. In G. B. Doern & B. B. Purchase (Eds.), *Canada at risk? Canadian public policy in the 1990s* (pp. 46-58). Toronto, ON: C. D. Howe Institute.

Soja, E. (1991). Poles apart: Urban restructuring in New York and Los Angeles. In J. H. Mollenkopf & M. Castells (Eds.), *Dual city: Restructuring New York* (pp. 359-376). New York: Russell Sage.

Stanback, T. M., & Noyelle, T. J. (1982). *Cities in transition: Changing job structures in Atlanta, Denver, Buffalo, Phoenix, Columbus, Nashville, Charlotte.* Totowa, NJ: Allanheld-Osman.

Statistics Canada. (1975). *1971 census of Canada: Labor force and individual income: Employment income for Canada, provinces, and census divisions* (Catalogue 94-711, Vol. 3, Pt. 1). Ottawa, ON: Minister of Supply and Services.

Statistics Canada. (1983a). *Census Tracts: Selected social and economic characteristics: Montreal* (Catalogue 95-962, Vol. 3, profile series B). Ottawa, ON: Minister of Supply and Services.

Statistics Canada. (1983b). *Census tracts: Selected social and economic characteristics: Toronto* (Catalogue 95-977, Vol. 3, profile series B). Ottawa, ON: Minister of Supply and Services.

Statistics Canada. (1983c). *Census tracts: Selected social and economic characteristics: Vancouver* (Catalogue 95-979, Vol. 3, profile series B). Ottawa, ON: Minister of Supply and Services.

Statistics Canada. (1991). *Canadian economic observer: Historical statistical supplement.* Ottawa, ON: Minister of Industry, Science, and Technology.

Statistics Canada. (1993a). [Number of wage earners 15 years of age and over who worked in 1990, by wage groups, industry groups, mother tongue, and official language: Metropolitan Montreal]. Unpublished special tabulation.

Statistics Canada. (1993b). *Selected income statistics: The nation.* Ottawa, ON: Minister of Supply and Services.

Uchitelle, L. (1994, July 25). U.S. corporations expanding abroad at quicker pace. *New York Times,* p. 1.

U.S. Bureau of the Census. (1973). *1970 census of population: Subject reports: Journey to work: Characteristics of workers in metropolitan areas.* Washington, DC: Government Printing Office.

U.S. Bureau of the Census. (1984). *1980 census of population: Subject reports: Journey to work: Characteristics of workers in metropolitan areas.* Washington, DC: Government Printing Office.

U.S. Department of Transportation, Bureau of Transportation Statistics. (1994). *1990 census transportation planning package: Statewide CTPP* [CD-ROM]. Washington, DC: Bureau of Transportation Statistics.

5 Globalizing Economies and Cities: A View From Mexico

DANIEL HIERNAUX

The changes observed as the world reaches the end of a millennium cannot be explained as the continuation of great tendencies seen in the past. In effect, after the mid-1970s, the bonanza years have died away, making way for a restructuring of global economic systems.[1] Restricting the analysis of these changes to the sphere of international relations, however, is not sufficient: New tendencies arise (new potentialities and new contradictions) that challenge not only political relations but also the technological paradigm that guided capitalist societies for decades, the bases of social ethics, and the organization of world space.

The changes that have caused such upheavals in modern societies lead one to believe that far from having found equilibrium, an end of history, or the unification of contradictory tendencies in a unique model of world society, present-day social systems struggle amid contradictions; social scientists search for regularities and hope to recognize processes or new perspectives that may be more generally applied (Friedman, 1993; Vaillancourt-Rosenau & Bredemeier, 1993). This convulsed milieu—this new era of international cities—is obviously a time of upheaval, of contradictory perspectives and tendencies, and of challenges and opportunities.

In the present world scene, one aspect has caught the eye of many social scientists: the growing role of territorial affairs in explaining large-scale world processes (see, e.g., Knight & Gappert, 1989). The relationship between these processes and territory can be studied from two complementary approaches: On the one hand, the global transformations of economies and societies no doubt reshape territories (at different scales), but, on the other hand, present territorial configurations act as a powerful motor, a key variable in global transformations. Rather than the importance of territoriality diminishing as a result of globalization due to a "spatial expansion," today a revitalization of territorial issues is occurring, starting with global changes experienced on a planetary level.

This chapter will highlight the role played by big cities. Although some strictly maintain local or regional managerial functions, others add or improve a new international dimension to their activities, as elements of articulation on a planetary scale within a globalized economic system. I will addresses this issue, setting aside strictly regional aspects or those concerning the internal reorganization of national territories vis-à-vis the changes mentioned. Further, this chapter emphasizes a Mexican point of view, necessary to understand the territorial logic that goes beyond the frontiers separating the three countries that are in the process of developing a continental relationship on the basis of a North American globalization. Certainly, this outlook differs from that of the United States or Canada; it is obvious, however, that globalization cannot exist without an integral interpretation that includes the views of these three national components on the accelerated globalization process.

Globalizing Economies, Globalizing Territory

The title of this section may seem obvious to many; it is not, however, a generalized outlook among decision makers in Mexico, who analyze globalization from innovative socioeconomic and political perspectives yet continue to produce a traditional discourse regarding national territory, especially regarding the city system (see Secretaría de Desarrollo Social [SEDESO], 1993). Thus, I will mention some of the conceptual bases used in analyzing the Mexican situation.

First, in all countries, changes in the relations between partial spaces of a national territory have occurred as a result of globalization processes. The Mexican experience shows that these changes are of different types:

1. A progressive disintegration of the so-called city system or, at least, an accelerated recomposition of relationships, hierarchies, levels, and so forth, is the result of the prevalent economic model no longer being exclusively domestically oriented (import substitution) but directed toward other countries as well.[2] Some cities, particularly those that are more fully integrated into the international economy, show a great capacity for demographic and economic growth and for the modernization of their physical structures.

2. These transformations have been relatively quick, and they contradict a good deal of past discourse regarding the stability of territorial forms through time.[3] The speed of these territorial changes must be assimilated at the same rate as the changes that are produced in other fields, among them communications and network systems, as will later be seen.

3. As relationships between cities are redefined, new infrastructures are needed to transport not only merchandise but also people. One of the keys to understanding the new national and global spaces lies in the connections between these.[4]

4. Not all localities and spaces are able to become part of this model; thus, new forms of social and regional inequalities appear that affect both cities and regions. New rules of the game among cities are evident: market forces, at least in cities in the United States and Canada. This is less so in Mexico, whose cities have been strongly marked by state policies and by a centralized and territorially concentrating economy. Thus, the "competitive advantages of the cities" vis-à-vis the global system have become a central element in understanding their internal functioning and their ability to become an integral part of an international system of cities.

Second, new forms of economic management have arisen, the result of the existence and progressive imposition of a new technological paradigm. This has allowed the production process to become fragmented and has established "remote control" management, keeping complete control over diversified and partial processes disseminated in multiple but articulated spaces.[5]

Thus, the introduction of a reticular space covers the whole world, generates new articulations, and takes advantage of the entire planet within a context of a new production-management-distribution model.[6] This reticular space has various components. On one hand, there are concentration nodes of production and management activities and material or nonmaterial networks along which decisions, capital, people, information, and products flow. Obviously, new power structures also flow along these networks, and new confrontations among actors in the global system arise. The planet enters an age of great nerve centers, of management nodes in the world space. Is this the era of great international cities? Certainly it is, from the moment when the relationship between the great centers of activity and of urbanization is evident. Nevertheless, not all activities are carried out in large cities, megalopolises, or great metropolitan areas; they can be observed in other situations: areas of secondary production (areas of outbound plants, export zones, etc.) and small concentrations of great power in the world system (technopoles, for example; see Benko, 1990).

Therefore, the era of the internationalization of cities is not necessarily and exclusively the one of megacities but the inevitable dominion of those cities able to place themselves at an international level and to develop functions that prove their ability to internationalize. This is important in the Mexican case, in which Mexico City seemed to benefit from the privilege of internationalization as regards a domestic market model but now faces a strong resistance to its dominion in the globalization era.

Another relevant observation (not only for the Mexican case) is that the internationalization of cities does not imply that all their territory and population are capable of entering the process. With the fragmentation of urban spaces and the existence of internal areas of high modernization and internalization are the presence and even the growth of areas of noticeable economic stagnation, increasing poverty, and advanced marginalization. (For the Mexico City case, see Hiernaux, 1995; for the São Paulo case, see Santos, 1990.)

The increasing presence of new demands for autonomy from smaller spaces, a question no doubt more relevant for those who study municipal issues, is latent from the moment when the city systems, hierarchies, and legitimacies are redressed. In the Mexican case, it can be proven that a clear relation exists between this situation and present political change and demands.

Globalization From a Mexican Outlook: Previous Economic and Territorial Bases

The Mexican case differs from others because its current economic, political, and territorial processes are far removed from those in the past. Without going into a detailed account of the characteristics of the model that preceded the liberalization process, I will highlight some of the aspects of the Mexican system to better explain the potential of Mexico's cities in relation to economic globalization and urban internationalization.

In the first place, the capital's dominance over the national urban system is characteristic of the Mexican territorial system. Without mentioning the well-known pre-Colombian concentration, the *Porfiriato*[7] shows that Mexico's first great modernization and first attempt at following in the steps of developed countries left a clear-cut mark of centralism at all levels and a city system that reinforced the concentrating guidelines that stemmed from the capital. The 1910 revolution destroyed a great deal of the country's economic capacity, disintegrated the country's limited communications system, but left Mexico City intact. Thus, the stabilization that followed Calles's (1924-1928) and especially Cárdenas's (1934-1940) administrations rebuilt and extended the infrastructure system but favored the territorial, economic, and political centralism that characterized Mexico until well into the 1970s.[8]

During the 1970s, experts on Mexico emphasized this marked concentration around Mexico City, and not without reason: Significant percentages of industrial activity, tertiary functions, and population had centered around the nation's capital (see Garza, 1985).

It follows that Mexico's first international city was the capital: It concentrated banks and international financial transactions, it became the site of multinationals, it attracted a greater foreign population than any other city, it developed a culture that combined the indigenous with the modern international, and it copied architectonic models. In sum, it became the epicenter of international modernity.[9]

Mexico City was not the only city that had relations with foreign countries, however. Examples include Tijuana's relationship to the state of California during the 1920s, especially during Prohibition, and Acapulco, which became an internationally favored tourist attraction after Batista was overthrown in Cuba in 1958. During the three decades of development based on import substitution, Mexico City did not have a monopoly over the country's international relations, but it was the only city that achieved a more or less complete international profile.[10] Mexico City's internationalization must be understood as a monopoly associated with its predominance in the Mexican economic, political, and territorial system. This supremacy was reinforced through the years because of the difficulty other cities faced in achieving a key role in the economy, the increasing articulation between national and foreign capital within the peculiar protectionist-liberalization mode that characterized the Mexican economy during those years, and the foreign capital that settled in Mexico City even when many of its operations were handled outside the city (as was the case with mining, for example).

It cannot be denied that Mexico City became a great city of demographic and economic concentration known to all: From a mere 1.6 million inhabitants in 1940, Mexico City increased to 5.1 million in 1960, 8.6 million in 1970, 14.1 million in 1980, and, finally, 15.2 million in 1990. Nevertheless, many analysts have sinned by exaggeration, because other cities also developed important functions even before 1980. Monterrey, Guadalajara, and, to a certain degree, Puebla are cases in point: They became new bases on which the economy was shaped and were gradually transformed into metropolitan areas. Clearly, those three cities developed their demographic position and economic base through their role as macroregional centers (Guadalajara for the northwest, Monterrey for the northeast, and Puebla for the south and southeast [Yucatán Peninsula]. Their economic base allowed these cities to develop a local and regional market and to substitute, to a relative degree, the centralized market of Mexico city. Without a doubt, Monterrey is the only city that established more relations with the United States because of its strategic location close to the border, its limited relations with central Mexico, and the autonomy of its entrepreneurs. Since the beginning of this century, steel fabrication, beer brewing, and glass production have been among the central activities of this city, with some trade relations with Texas.

As to the internationalization of these cities, they were able to develop only partial relations with other countries. Most of the more important relations between foreign economies and Mexico were channeled through the capital.[11] Because of the existence of a highly centralized model of economic and political relations, the nation's capital appropriated not only the central offices of foreign firms and banks but international commercial and cultural relations as well. When these centralized relations are replaced by market forces, as is the case today, it is clear that international functions can be modified and relocated to include cities outside the capital.

The stage characterized by the oil boom seemed to reinforce the ability of other Mexican regions to "win space" in the model and shake off centralization. A significant redistribution of public investment took place during this period, particularly regarding basic industries, infrastructure, and some support to local development. An associated demographic redistribution also occurred, mainly as a result of oil extraction and refining and petrochemical activities. But if those activities and the related demographic changes were the beginning of the actual process, the inertial concentration effect was still so high that no clear tendency of significant economic and demographic relocation can be perceived in the 1980 population and economical census—only some shift and signs of further movements. The Mexican government also tried to develop large sectorial and regional investment projects, among them iron and steel industries in Lázaro Cárdenas (in the state of Michoacán) and Monclova (Coahuila); copper in Nacozari de García (Sonora); oil in various places in the southeast; and international tourism in Cancún (Quintana Roo), Ixtapa (Guerrero), and

Loreto and San José del Cabo (Baja California Sur). (For an overview, see Hiernaux, 1989.)

Although these projects brought about a territorial redistribution of the economy and the population, their results were limited. They also did not significantly modify the centralist scheme, despite, for example, Cancún being an internationally renowned tourist attraction. Nevertheless, Cancún has around 200,000 inhabitants and Lázaro Cárdenas some 80,000. This represents a considerable regional change as new poles arose, but the national impact is relatively small compared with the scale of Mexico City. In fact, not until 1985 was Mexico able to shake the protectionist state policies that had been in effect for close to 40 years because of (a) an international conjuncture that favored less state intervention, (b) pressure from Mexico's creditors, (c) lack of fresh oil resources after the international fall in prices in 1981, and (d) economic stagnation between 1982 and 1985 that forced the implementation of new commercial policies.

The most important decisions regarding the direction for the Mexican economy and society were made after 1988, under the Salinas administration. Since 1983, however, several important measures were adopted that cleared the way for the more recent changes: (a) the foreign debt was restructured; (b) the economy was gradually opened; (c) industries in nonstrategic sectors were gradually privatized, albeit slowly; and (d) relations between the government, Mexican entrepreneurs, and international economic and political sectors were reconstructed in a climate of mutual trust.

After 1988, these measures were carried still further, and the changes came about more rapidly. Many sectors that previously had not shown significant dynamism or that might slow down the Salinas modernization project were given new directions. For example, in agriculture, a constitutional amendment was made regarding the postrevolutionary system of *ejido* (peasant property under state control, impossible to sell or rent); various traditionally state-controlled sectors (petrochemicals, transportation, telecommunications, urban works, etc.) were privatized and gradually opened; and changes were brought about in the political arena (victories of the opposition at the state and municipalities level). Worthy of note are the amendments regarding foreign investments and, above all, a change in attitude toward the United States and Canada. This change has affected the opposition party, which had been willing to resort to these countries to contest official Partido Revolucionario Internacional (PRI) policies.

In territorial terms, important changes have come about. I will discuss this issue in the following section to determine if there have been geographical winners and losers.

Winners and Losers: General Hypotheses

Despite the official discourse, few harbor any doubts about whether the globalization process encourages a game between winners and losers; nobody is naive enough to think it is a "zero-sum" score. There are winners and losers among economic sectors as well as societal and economic agents. This has led to a debate within Mexican society between those who believe that the profits are greater than the losses and those who believe the contrary. I do not wish to enter into this complex and argumentatively rich discussion here but rather to reconstruct the development of a game that has yet to be concluded—a game in which none can doubt that there will be regions and cities that win and others that lose.

What are the variables that affect profits and losses? To begin with, I must point to the unequal distribution of traditional job losses in the Mexican economy. The older urbanized and industrialized areas are going through difficult times because of the continuous closing of companies. Reduced productivity, obsolete technology, inability to restructure the market, and lack of innovations in product design are some of the problems that can be identified in industries or sectors that end up as losers.

Together with the global concentration previously mentioned, there has also been a strong concentration of the smaller productive units to face such brutal changes. As a result, the comparison between cities and regions is convincing. In 1982, all big cities suffered from an unprecedented recession; but a few years later, it became clear that the cities that would lose the most jobs were those that had traditional, protected, and relatively inefficient industries.

The globalization of the economy would then have a differential effect on cities and regions. At the risk of oversimplifying, the following can be stated as hypotheses (for a more complete analysis, see Hiernaux, 1994):

1. Northern regions, which had already been opened as a result of their traditional limited relations with the rest of the country, their proximity to the U.S. border, and the constitution of an artificial liberalization statute within a protectionist scheme (laws on duty-free zone laws and the *maquiladoras*), will profit the most from the economic liberalization. Although *maquiladoras* provide many jobs, such employment is very volatile: In recent months, under the pressure of competition and possibly because of political fears, important plants have closed. The urban economy under a *maquiladora* domination is therefore unstable.

2. Some medium-sized cities not far from the border may benefit from the liberalization process as assembly plants for the automobile industry are established in the area. Chihuahua, Hermosillo, and Saltillo are the midscale cities in this situation. Because of a much more long-term industrial location under the rules of the automotive industry in the North American Free Trade Agreement (NAFTA), clearly these urban economies are reinforced and are made more stable. At any rate, they do not develop the same demographic base as the border cities, whose demographic growth is related to job creation but also, and perhaps mainly, to the migration effect.

3. States in central Mexico are facing greater difficulties as they try to redirect their economies because of the overwhelming weight of traditional industry and its particular way of operating. In particular, the metropolitan zone of Mexico City is losing jobs, with some relocating in the central region (seven states around the capital). Because no study has been conducted on this process, it is difficult to advance data about job loss and the countereffect of relocation.

4. For the southern regions, except for a few isolated locations that can easily become part of the international economy thanks to certain specific competitive advantages (tourist sites, for example), most of the territories are excluded from economic integration. The Chiapas conflict that started January 1, 1994 is the result not only of traditional marginality but also of the perception (expressed by Subcomandante Marcos) that the economy was leaving the people behind.

Practice has revealed something similar to these hypotheses that were put forth a few years ago; however, we need both more case studies and an aggregate study that can verify the hypotheses with general data. Beginning with the 1989 economic census, relevant data can fit into a framework, but even so, the rate at which changes have been taking place since then is such that many variations were not registered in the censuses.[12] Among the topics that need to be developed are those related to the trade relations between particular cities and states and the rest of the world. Fortunately, recent data series are now available (but not broadly diffused). Yearly series for 1990 to 1993 on imports and exports by state and 99 groups of products will allow the development of a statistical analysis of the pattern of international commerce. A preliminary overview shows the emergence of new export zones, particularly regarding some products with clear ties to imports, such as cement. Beverage (beer) and automotive (e.g., motors and complete car units) exports are also composing a higher proportion of the actual export amount. But on the other side, Mexico City is still reported to export products that are not produced locally because its numerous headquarters report their operations there.[13]

At this time, I cannot extend the analysis on the regional dimensions of profits and losses. I can point out, however, that regional losses raise serious problems for the government, which, with funding from the World Bank, has begun an extensive program to address the needs of four impoverished southern states (Oaxaca, Guerrero, Chiapas, and Veracruz) that are not able to jump on the "train of modernity."

The Role of the Big Cities: Antagonism, New Relations, and the Emergence of a New Internationalized System

In this final section, I will analyze the central theme of this chapter: the role that cities play in the globalization process and, particularly, whether Mexico will be able to launch one or more cities into the narrow Olympus of international cities.[14] Once more, the starting point will be the beginning of the 1980s, particularly after the fall in oil prices. The two cities that benefited most from the oil boom, directly or indirectly, were Monterrey and Mexico City.

Monterrey profited because it obtained numerous contracts and because its entrepreneurial group became firmly established and was able to maintain an important profile in all the works that were carried out by the López Portillo administration. But the Monterrey group's consolidation dates further back; its history has been amply documented, and no doubt it is the most successful example of urban development with the support but only a minimal presence of the state. Monterrey is noted for its entrepreneurial standing, the modern behavior displayed by its employers, the almost nonexistent labor unions, and the permanent presence of the United States, only a few hours away. The oil boom, coupled with these antecedents and the rapid expansion of the city's industrial groups, however, almost led to its complete fall after 1982. Only the support offered by the state—through the recently nationalized banks—saved Monterrey from complete bankruptcy. The Mexican state flew to rescue its best industrial bulwark with a historically unprecedented loan.

Once they recovered from their stupor, the entrepreneurs restructured their firms, a process that led to greater economic transnationalization and to a search for new markets and an aggressive strategy to expand their activities.[15] This led to a reexamination of state-government relations; today the PRI government administration has strong ties with business, an unusual case in Mexico. It is common knowledge that the office of the presidency discusses many important economic decisions with business groups, a situation that even in Mexico is looked on unfavorably not only by the opposition but also by wide traditional sectors of the official party (Villareal, 1993).

Mexico City's economy is another matter altogether. Until 1985, its attitudes had not greatly changed. The earthquakes that shook the city that year also mobilized civil society, making natural dangers and pollution more evident than ever before, although they did not create an awareness of the capital's economy. Because of its size, the diversification of its economic interests, and its highly politicized management, among other reasons, Mexico City is not aware of the need to establish bases for metropolitan economic policies. It is only in the last 2 years, probably as a result of the growing social repudiation of industry closings and the contraction of the labor market, that not only political management per se but also economic groups and associations have a newfound interest in the city's economy. At the same time, opposition parties have begun to point to the decay of Mexico City's economy and even to use it as part of their campaign slogans. All this has happened after serious damage had already been done. Census data in 1989 show that Mexico City had suffered important job losses, which were magnified in certain sectors such as the metallurgical industry, one of the bastions of import substitution. The situation has only become worse since then. On the other hand, the informal economy has surged, especially street sales in the capital's historic center. All these elements have begun to sound the alarm regarding the economic possibilities of the metropolitan area (see Hiernaux, 1993b).

The central areas have lost more jobs, especially in traditional industrial branches, than other parts of the city. These jobs were replaced by a growing tertiary sector. The conurbation formed by municipalities that are close to the old industrialized areas of the Federal District suffer from the same situation. In addition, all this goes hand in hand with a marked demographic contraction in central districts, which have lost a good deal of their population, including most of their youths. The municipalities that are further from Mexico City's metropolitan area, however, show a growth of their industrial plant. Despite the lack of statistical data, I hypothesize that this growth is a result of the relocation of industries in industrial parks far from the city or in smaller localities of what can be called a metropolitan region.

From another point of view, it is clear that Mexico City is recovering a vigorous presence in international functions because many foreign firms continue to establish their central offices there. An important share of the 500 largest Mexican companies have located in or around Mexico City, reflecting the city's growing role in management, to the detriment of other economic spaces that reap the benefits of only the units of production. This situation enables the capital to reduce pollution, but it is evident that management functions of this type cannot absorb a significant percentage of the population that demands jobs. On the other hand, it is obvious that these headquarters, in their turn, demand services and activities; thus, I have been able to observe, if not measure, an important increase in banking activities in the city, an expansion of the Mexican Stock Exchange's activities, a profusion of small offices and service activities, and diverse businesses articulated with the management of the industrial process.

Thus, Mexico City is able to reconstruct an employment base under a new formula: tertiary activities. The results are plain to see; in addition to significant retractions of the population in old industrialization areas, the city does not have a long-term economic plan to reconvert its work plant and direct it toward other activities. The only prospects that remain are unemployment, underemployment, and emigration, this last having reached significant levels.

Mexico City's international functions have expanded to such a point that a large percentage of high-income Mexicans live in the capital. As a result, the market for luxury goods, international services, and social and cultural activities "worthy of the First World" are in constant demand by the well-to-do. Those international functions can be reviewed following the central ideas of Soldatos's (1989) list: more foreign investment; banks trying to assault the still largely closed financial market; a growing stock exchange classified between the new emerging places in the world; growing role of international technology, participation of foreign technicians and consulting groups; a much more internationalized profile of cultural activities, concerts, and economic fairs; and increased investments from Canada and the United States in malls, office buildings, and so forth. Even higher education is showing that influence with a growing market of international joint operations between Mexican and external universities. The international airport was substantially developed with more than 100,000 square meters of new buildings to house the new airlines offices and support the growing traffic.

In sum, although the federal government has ceased to exercise certain central functions, many of the international ones are still in Mexico City.

These include, for example, the NAFTA negotiating group, consulting groups, companies that favor the country's economic liberalization and that encourage foreign capital, foreign commerce banks, and a recently opened world trade center.

What possibilities do other cities have of becoming part of this process? In other words, which are the required conditions and perspectives for the constitution of a set of international cities in Mexico? My view is relatively pessimistic because Mexico City's centrality has not in the least diminished. As in all developing countries with an age-old centralist tendency, municipal and local power is a rarity, despite recent reforms.

It is well known that even in recent years, cities such as Guadalajara have not been able to deal with urban emergencies such as the pipeline explosion caused by oil wastes in 1992. Monterrey faces serious pollution, and Puebla has not been able to change its status as a provincial city, despite having a population of more than 1 million. Clearly, centrality is not the exclusive result of an economic model but the product of a given concept of political, social, and cultural life and, ultimately, of a concept of territory. Therefore, the globalization of the economy cannot guarantee the globalization of the cities, except those that already wielded a certain amount of control over past economic processes and, for those same reasons, began to take shape as the cities with the greatest potential on the national level. It is impossible to name one midsized city that has yet been able to go beyond the profile established for a city open to foreign investments and that has managed to establish itself as the seat of other important international functions. It is true that some cities have acquired certain characteristics of international activity: Cancún and Acapulco, among others, have developed an important tourist image. But in all cases, the activities are secondary in the world economy and the cities are essentially monofunctional. The same can be said of the oil and mining cities and others that boast activities related to the global economy.

Without a doubt, one of the key elements for a city's internationalization is the establishment of a specific, strategic role, defined on the basis of geographic location, infrastructure conditions, or the predominance of functions that are vital to the world economy in a given era. No Mexican city except the capital has been able to develop these conditions and at the same time maintain continuous growth with a firmly established domestic base and receive the corresponding recognition. International functions, then, must appear or become consolidated on the basis of a well-defined domestic policy. This seems to have been the case in some cities that achieved international status by means of an initiative directed toward the consolidation of educational, cultural, research, and international linkage functions or to a more spontaneous process that was nevertheless reclaimed and encouraged by groups in power (not exclusively the government). Nevertheless, the difficulties Seville (Spain) faces, after the world fair and the huge investments, prove that geographic location, conditions of the domestic economy, and the perspectives for entering the global economy cannot be changed by public will alone. The large investment projects in Latin America illustrate the ineffectiveness of governmental promotional measures. Therefore, some cities will have advantages over others, but it is necessary to find the strategic profile for international integration and the appropriate means to encourage inclusion in each given case. Competitive advantages are built through time; they lead to the construction of competitive spaces. History, especially flourishing examples such as Venice or Hong Kong, can be a great help in understanding the real mechanisms that lie behind a successful international city.

For Mexico, it is evident that a laissez-faire policy would be completely erroneous. In effect, competition among cities is exacerbated and can be defined on two levels: competition between Mexican and foreign cities and among Mexican cities themselves. On the first level, Mexican cities have comparative lags in infrastructure and in the conditions of their social and economic life that place them at a disadvantage. The recent liberalization of its economy and the flexibility of its urban policies, however, have shown that even Mexico City, with its age-old defects, is able to offer appropriate sites for international functions. To this effect, and despite the absence of an explicit global policy, the city's government has undertaken a policy of partial modernization projects in traditional areas and of the promotion of new areas of modern development. Mexico City has areas on a par with those in any highly developed industrialized nation, with similar living conditions for those few who can afford them: the best schools and universities, top-level medical care, exhibits and other cultural activities, and residences of enviable quality.

There are, however, structural limits to all of this: "socialized" problems left over from a Fordist city, such as pollution, lack of services, chaotic traffic, considerable unemployment, and menacing poverty. To a certain extent, price increases in Mexico City propitiated its duality because the poor were pushed toward a periphery that is physically and socially further removed. But this does not completely eliminate the latent threat, and the riots in Caracas (Venezuela) and supermarket holdups in Brazil are not far from the minds of the politicians. The redistribution policy undertaken by the National Solidarity Program does not dispel the social risks; it only lessens them and reduces the political pressure on the current economic model.

On the other hand, in addition to these structural limitations and as a result of deficiencies in the distribution of income that were inherited from the past, it has become evident that international functions tend to branch out to other cities. Not all of these have competitive advantages, as already pointed out, but a few cities have managed to take off, particularly Monterrey. In this case, the business-local government relationship is evident as a key element of the city's success in its new world relationship. If its integration in the global economy was forced by the crisis, the search for a better position as a city is the result of a restructuring of entrepreneurial positions and of the public will. The city's government has signed brotherhood agreements with numerous cities on the international scene and is preparing itself to carry out the international functions of the city. It seems that the ability to participate in the era of international cities, in the Mexican case, has to do with a strong integration of a city's dynamic sectors and, mainly, with the formulation of concrete policies by urban authorities in close association with entrepreneurial groups.

In this scenario, Monterrey has shown great progress in developing international-level activities, not only on the economic side but also with the start of a new cultural center policy with considerable private investment. Mexico City, however, is still situated on the dangerous threshold of governmental populism, concentrated opposition, and, even more critical, a lack of definition of its economic model.[16] Neverthless, its legacy as a central city in control of international processes, which it has enjoyed for centuries, gives it a definite advantage with respect to its image. And, within the context of this rivalry, we can not discount other smaller, less dynamic cities, such as Puebla or Guadalajara, as possible competitors in this era of international cities.

NOTES

1. The bonanza years include, from my point of view, the oil boom years between 1978 and 1982. The price of oil dropped in April 1981, and the Mexican economy started a deep recession. But from a broader point of view, the large imports substitution phase gave a continuous and substantial growth to the Mexican economy (around 9%) from the 1950s until the end of the 1960s.

2. Recently, Krugman and Livas Elizondo (1992) stressed that question from a traditional and modelistic aspect but not without a strong tie with reality.

3. Exaggerated demographic projections from the 1970s and early 1980s claimed that Mexico City's population would be more than 30 million inhabitants by the year 2000. Actually, 22 million seems a more realistic estimate because the growth rate of the total population of metropolitan Mexico is around 1% yearly.

4. During the last years, the question of connections has been central in the economic policy: A new expressway program has been proposed under concession; ports have been privatized; a new port and harbor law and port administration system is under study; a policy of free sky has been implemented; and, obviously, the Mexican Telephone Company has been restructured and privatized.

5. Actually, a large bibliography arose about those questions, particularly from the view of the regulation school in economy. Of interest is the critical article of Amin and Robins (1990) against that tendency of a new idealist geography represented by, for example, Benko (1992), Benko and Lipietz (1992), Scott (1988), and Storper and Scott (1992).

6. That means a new concept of time-space relations, such as those expressed in the modern stage of capitalism. See Kern (1983).

7. The *Porfiriato* corresponds to the period from 1875 to 1910 when General Porfirio Díaz acted as dictator in Mexico. This period was characterized by the transfer of indigenous properties to private hands; the arrival of external capital in railroads, electricity, and industrial activities; and a general ambiance of repression and lack of free elections that culminated in the revolution of 1910.

8. Nevertheless, the intense land distribution that took place during the Cárdenas period reduced temporarily the urban concentration that had strongly increased as a result of the civil war.

9. For example, the California style for residences was commonly used during the presidency of Miguel Aleman (1946-1952).

10. See Soldatos (1989) for definition and analysis of international urban functions; see also Fry, Radebaugh, and Soldatos (1989).

11. A significant example is the location of a Volkswagen plant in Mexico City, which precluded an international prospect for the city of Puebla.

12. For a discussion of the data shortages and of the necessity of gaining additional data, see Hiernaux (1993a).

13. See, for example, the data series presented each August by the journal *Expansion* on the 500 biggest enterprises of Mexico.

14. The following paragraphs are a synthesis of my project at the Universidad Autónoma Metropolitana-Xochimilco.

15. Note, for example, the aggressive behavior of Cementos Mexicanos when it recently bought a Spanish cement plant. Transnationalization was implicitly allowed by offering part of the assets to international financial groups to lower the debt.

16. The former functionary (*regente*) in charge of Mexico City was one of the potential candidates for the official party in the August 1994 presidential elections. Therefore, his interests were less oriented to the economic future of Mexico City. Because he is actually out of Mexican politics, the city was taken in charge by a low-profile mayor, who had no real intent to keep the position, until December 1, when the government changed.

REFERENCES

Amin, A., & Robins, K. (1990). The re-emergence of regional economies? The mythical geography of flexible accumulation. *Environment and Planning D: Society and Space, 8,* 7-34.

Benko, G. (1990). *La géographie des technopoles* [The geography of technopoles]. Paris: Masson.

Benko, G. (1992). Espace industriel, logique de localisation et développement régional [Industrial space, locational logic and regional development; in special issue on economic restructurations and territories]. *Espaces et Sociétés, 66-67,* 129-146.

Benko, G., & Lipietz, A. (1992). *Les régions qui gagnent: Districts et réseaux: Les nouveaux paradigmes de la géographie économique* [The winning regions: Districts and networks: The new paradigms of economic geography] (Economie en Liberté series). Paris: PUF.

Friedman, J. (1993). Order and disorder in global system: A sketch. *Social Research, 60*(2), 205-234.

Fry, E. H., Radebaugh, L. H., & Soldatos, P. (Eds.). (1989). *The new international cities era: The global activities of North American municipal governments.* Provo, UT: Brigham Young University, David M. Kennedy Center for International Studies.

Garza, G. (1985). *El proceso de industrialización de la ciudad de Mexico 1821-1970* [The industrialization process of Mexico City, 1821-1970]. Mexico City: El Colegio de México.

Hiernaux, D. (1989). Grandes proyectos de inversión y desarrollo regional [Large investment projects and regional development]. In F. Brunstein, E. Laurelli, A. Rofman, & A. Vidal (Eds.), *Grandes inversiones publicas y espacio regional: Experiencias en América Latina* [Large public investments and regional space: Experiences in Latin America] (pp. 337-370). Buenos Aires, Argentina: CEUR.

Hiernaux, D. (1993a). *Cambio económico, liberalización y la información para el estudio regional-urbano binacional* [Economic change, commerce liberalization and information for binational urban and regional studies]. Paper presented at the Binational Census and Data Conference organized by the University of Texas at Austin and the Association of Borderland Scholars, El Paso, TX.

Hiernaux, D. (1993b). La ciudad de Mexico frente a los cambios económicos: Las nuevas perspectivas de la apertura [Mexico City facing economic changes: New perspectives from the liberalization process]. In A. Bolivar, R. Coulomb, & C. Muñoz (Eds.), *Metrópoli, globalidad y modernización* [Metropolis, globalization and modernization] (pp. 205-240). Mexico City: Universidad Autónoma Metropolitana and Facultad Latinoamericana de Ciencias Sociales (FLACSO).

Hiernaux, D. (1994). De frente a la modernización: ¿Hacia una nueva geografía de Mexico? [Facing the modernization process: Toward a new geography of Mexico?] In M. Bassols (Ed.), *Campo y ciudad en una era de transición* [City and countryside in a transitional era] (pp. 19-46). Mexico City: Universidad Autónoma Metropolitana-Iztapalapa.

Hiernaux, D. (1995). *Nueva periferia, vieja metrópoli: El valle de Chalco, México* [New periphery, old metropolis: The Chalco Valley, Mexico City]. Mexico City: Universidad Autónoma Metropolitana-Xochimilco; Toluca: Gobierno del Estado de México.

Kern, S. (1983). *The culture of time and space, 1880-1918.* Cambridge, MA: Harvard University Press.

Knight, R. V., & Gappert, G. (Eds.). (1989). *Cities in a global society* (Urban Affairs Annual Review, Vol. 35). Newbury Park, CA: Sage.

Krugman, P., & Livas Elizondo, R. (1992). *Trade policy and the third world metropolis* (Occasional paper). Cambridge, MA: National Bureau of Economic Research.

Santos, M. (1990). *Metrópole corporativa fragmentada, o caso de São Paulo* [Corporate and fragmented metropolis: The São Paulo case]. São Paulo, Brazil: Nobel Editores.

Scott, A. J. (1988). Flexible production systems and regional development: The rise of new industrial spaces in North America and Western Europe. *International Journal of Urban and Regional Research, 12*(2), 171-186.

Secretaría de Desarrollo Social (SEDESO). (1993). *Programa de 100 ciudades* [100 cities program; mimeo]. Mexico City: Author.

Soldatos, P. (1989). Atlanta and Boston in the new international cities era: Does age matter? In E. H. Fry, L. H. Radebaugh, & P. Soldatos (Eds.), *The new international cities era: The global activities of North American municipal governments* (pp. 37-72). Provo, UT: Brigham Young University, David M. Kennedy Center for International Studies.

Storper, M., & Scott, A. J. (1992). *Pathways to industrialization and regional development.* London: Routledge.

Vaillancourt-Rosenau, P., & Bredemeier, H. (1993). Modern and postmodern conceptions of social order. *Social Research, 60*(2), 337-362.

Villareal, D. (1993, September). *Cambios recientes en el proceso de industrialización en el area metropolitana de Monterrey y efectos a nivel regional* [Recent changes in the industrialization process in metropolitan Monterrey and regional impacts]. Unpublished manuscript, Toluca: Universidad Autónoma del Estado de México.

Part III

Municipal Networking and Intergovernmental Cooperation

6 Local Responses to Globalization and Regional Economic Integration

NORRIS C. CLEMENT

Introduction

Recent changes in the structure of global economic and strategic relationships are dramatically transforming the economic environment and presenting new challenges and opportunities for cities-regions.[1] This chapter synthesizes the major themes that have emerged in this context and references experiences in both Europe and North America as cities-regions respond to the new challenges and opportunities.

The first section integrates some of the main elements of recent thinking in the areas of industrial organization, regional development, and public affairs in the context of the increasingly internationalized economic environment, referred to here as the dynamics of globalization. During the last three decades, technological innovations have forced dramatic changes in the organizational structures of manufacturing and service firms as well as how and where they conduct business. These changes have led to important developments in the spatial distribution of both economic activity and employment patterns that, in turn, require governments at all levels to reassess their organizational structures and the functions they provide. The main effect of these changes has been to increase the responsibilities of city and regional governments in spearheading local economic development initiatives.[2]

The second section of the chapter outlines recent institutional responses to the increasingly internationalized, borderless environment in which these new responsibilities are assumed. Improved cooperation (i.e., alliances, networks, and partnerships) between all levels of government, business,

TABLE 6.1 Frame 1. The Change Process: Dominant Features (Technological Innovation)

Column 1		*Column 2*		*Column 3*
Transportation costs		Globalization		Competition
(Information)	→	(Production/distribution)	→	(+)
Communications costs		Decentralization		Cooperation
		(Geographical dispersal)		

and education/training/research institutions, together with the proper use of innovations in economic research and planning techniques, can increase the efficiency of both governments and private firms and enhance the competitiveness of regions. In the final section, I draw conclusions regarding cities' abilities to respond to the new sets of conditions facing them in the expanding internationalized environment.

The Dynamics of Globalization

This section synthesizes the main intellectual components of a paradigm shift regarding the new role of cities-regions in today's increasingly globalized economic environment. Frames 1 to 4 will illustrate four dimensions of change, moving from technological change (1) to the effects on economic structures (2), which in turn affect the changing spatial distribution of industry (3) and changing government structures and functions (4).

In Frame 1 (see Table 6.1), the main motor of change is identified as technological change, primarily in the transportation, communications, and information-computer sectors. These key services have become not only cheaper but also faster and more widely available throughout the world. Thus, many firms are now freer to go international—to geographically (re)locate any or all phases of production wherever costs are lowest or conditions are most appropriate to the firm's overall strategy.

"Smart firms" are now able to devise global strategies, decentralize and disperse their operations, and manage them from afar through fax, voice, and data transmission while shipping components and products via transportation systems that each day become cheaper and offer more options.[3] The resulting spatial redeployment of production, along with trade liberalization under multilateral institutions such as the General Agreement on Tariffs and Trade (GATT) and the formation of regional trade blocs such

TABLE 6.2 Frame 2. Economic Structures (How Firms Organize and Do Business)

Old/Traditional	→	*New/Emerging*
Agriculture/manufacturing	→	More service activities (information, financial, and tourism)
Large firms: economies of scale	→	Smaller firms: economies of scope (flexible production systems)
National market perspective (inputs and outputs)	→	International perspective • offshore sourcing (inputs) • production phases dispersed • global marketing (outputs)
Centralization of functions (everything in-house)	→	Decentralization of functions and blurring of manufacturing-service sectors results in: • increased outsourcing • growth of industrial clusters
Stable workforce (high pay)	→	More temporary, part-time workers
Hierarchical organizations	→	Work rules more flexible
Competitive activities (exclusively)	→	New strategies require cooperation, complex alliances, and networks (firms-governments-universities)

as the European Union (EU) and the North American Free Trade Agreement (NAFTA), as well as the emergence of many Third World countries as major manufacturing powers, have dramatically increased global competition. But as shown in Frame 1, the accelerated pace and cost of technological innovation has stimulated firms to increase their cooperation with other firms as well as with governments and universities.

Frame 2 (see Table 6.2) outlines the main changes in economic structure in recent decades. (The shift from old/traditional structures and institutions to new/emerging ones reads from left to right.) Well known is the transformation of production in all developed economies, although in varying degrees, from the real goods sectors (agriculture and manufacturing) to service- and information-based economies. The shift from large firms to small and midsized firms as the main source of new jobs is also well documented, as is the shift from a regional-national perspective to an international one (i.e., sourcing inputs—both component parts and human services—offshore and marketing outputs globally). Not so well known, however, are the enormous changes taking place within firms, especially high-tech companies:

TABLE 6.3 Frame 3. Regional Structure (Reflects Changing Location Determinants)

Old/Traditional	→	*New/Emerging*
Core-periphery: dynamic industries in core	→	Old core (restructured, diverse) New core (smaller, niche markets) Periphery (most other regions)
Location factors (costs: proximity to resources, markets, and suppliers)	→	Congestion costs in core rise Mature industries move to low-wage areas New industries move for climate, QOL, labor force, research facilities
Border regions in periphery	→	New core(?): new opportunities available but not guaranteed

- The emergence of computer-aided technologies that permit manufacturing firms to efficiently produce small quantities of custom-made products (flexible production systems responding to the economies of scope instead of the economies of scale associated with traditional, large manufacturing firms)
- The gradual demise of traditional assembly lines and hierarchical organizational structures with a reliance on highly skilled and experienced workers
- The greater use of outside technical and business services (outsourcing) and increased use of subcontractors, blurring the traditional distinction between manufacturing and services and creating industrial clusters[4]

Although these new practices often result in lower costs and an enhanced ability to respond to new market opportunities, employees' work roles and traditional notions of job security are changing considerably, increasing the need for worker retraining, counseling, and relocation—services usually provided by government.

Note that firms no longer are exclusively competitive. The need to innovate rapidly and the enormous costs of developing new products and techniques have pushed firms into cooperative networks and alliances with competing firms, governments, and university research centers (Reich, 1990).

Frame 3 (see Table 6.3) summarizes the effects of these changes on the regional (geographical) structure, or location of industry: the spatial dimension of the story. Traditionally, regional structures were viewed in a dichotomous framework of core-periphery. Traditional core areas—formed on the basis of proximity to resources, suppliers, and markets—were characterized mainly by large-scale manufacturing firms involved in the

TABLE 6.4 Frame 4. Governmental Structures and Functions (Responses to the Previous Frames)

Old/Traditional	→	*New/Emerging*
Stabilization policies: monetary/fiscal	→	Less national autonomy due to globalization, budget deficits
Industrial policies (national level)	→	(In U.S.) now expanding at national and regional levels (In EU) subordinated to regional policies shaped in Brussels
Social welfare system	→	Under attack everywhere as unemployment and welfare costs grow (immigrant role here important)
Large, centralized (national level)	→	Decentralized to local/regional (reinvent roles/structures) (EU) internal borders disappear transnational government emerges strategic urban networks form
Universities: national governments support education and basic research	→	Research and development, training functions at regional level increase as do alliances with firms and governments (private-public-academic partnerships)

production of producer goods and consumer durables as well as large financial firms. Meanwhile, activities such as agriculture, forestry, fishing, mining, and small-scale manufacturing activities were located in the periphery.

Now, however, it is clear that since industrial restructuring began in both the United States and Europe in the 1970s, traditional core areas have been transformed in both production techniques and product types (i.e., most mature products have been either automated or moved to low-cost regions). A new core has emerged, mainly in southern states. From a simple view, new core areas are characterized by all of the features of the modern firm, as portrayed in the previous frame. These firms, however, seek different location characteristics that, within some limits, emphasize qualitative aspects (e.g., climate and environment) over quantitative considerations. This, in turn, presents new opportunities for cities-regions in the old periphery.

Frame 4 (see Table 6.4) outlines the major changes in governmental structures and functions that have occurred mainly in response to the changes presented in Frames 1 to 3.

- At the national level, stabilization measures and monetary and fiscal policies, traditionally used to combat the extremes of the business cycle, have become less effective in the 1990s because (a) increased globalization results in fewer degrees of freedom,[5] (b) large budget deficits make stimulatory policies politically unacceptable, and (c) a growing awareness exists that economic problems in the United States and Europe are structural, not cyclical.
- Growing dissatisfaction with statist policies has resulted in a shift to the traditional conservative tendencies toward less reliance on and more selective uses of government.[6]
- Restructuring frequently is "good for the economy, bad for the people," at least in the short run, resulting in growing demand for government services and/or transfer payments. The attack on welfare has also meant scapegoating immigrants in many countries.
- As national governments attempt to reinvent government, new forms are tried, beginning with decentralization, expanding to strategic urban networks, especially in the EU (see below).
- With new emphasis on high-tech, high-value-added activity and rising income gaps between regions, "smart (subnational) governments" devise strategies to better play the new economic game by raising and improving or maintaining the quality of life. These include increased investment in physical and social infrastructure and the development of new relations between private, public, and academic institutions to improve productivity and the innovative milieu.

European and North American Responses to Globalization

Before examining cities' responses to the new conditions facing advanced economies today, it may be helpful to restate those conditions in summary form. In recent years, both economic and job growth have slowed, whereas economic, spatial, and administrative structures have undergone dramatic transformations. Thus, many see the decade of the 1990s as another watershed decade (as was the 1930s), marking the end of the post-World War II period. The 1990s, characterized by rapid economic and structural change, is driven mainly by technological innovation and an increasingly competitive international economic environment.

In response to rapid technological change and increased competition from low-wage countries, firms in advanced industrial countries have experienced a striking evolution in both structure and organization. Technological change has made it possible to operate on a global level with location of operations not tied exclusively to traditional cost considerations. In fact, for many firms, profits and market share depend more on rapid product development than simple cost reduction. Thus, it is increasingly important to locate in or create a milieu that supports innovation.[7]

These and many other elements have resulted in different spatial patterns of economic activity. The traditional core-periphery distribution has become more variegated, providing new economic opportunities for localities that previously found themselves condemned to the periphery.

In the meantime, traditional macroeconomic stabilization and structural change policies at the national level have become either ineffective or inappropriate for dealing with the increasingly globalized environment. Thus, many cities and states, mired in budget crises and a downward spiral of service cuts and tax increases, have responded by expanding their economic development efforts, mainly by (a) attempting to quickly improve the local business climate by reducing regulation, bureaucracy, and taxation; and (b) expanding city marketing efforts by encouraging local producers to increase exports and by attracting outside firms to locate and/or invest in their area.

This dual response is primarily a competitive one and frequently results in bidding wars in which, in exchange for jobs and investment, localities vie to provide the best package of incentives to prospective firms.[8] Other long-term, cooperative policies, however, are now employed in Europe and, to a lesser extent, in North America.

Urban Networking

This cooperative approach recognizes that although cities must compete in some areas (e.g., in attracting visible events such as the Olympic games and in luring investment), each city can increase its own competitiveness by cooperating with others in many other areas (e.g., exchanging information on programs to combat common urban problems and lobbying together for needed urban policies at the national or transnational level).[9] Urban networking has taken many forms and has been embraced enthusiastically by academics and public and private officials and practitioners (see "Urban Networking in Europe I," 1991; "Urban Networking in Europe II," 1992). It is easy to see why: Local officials in various cities working together to change or expand EU urban policies and to resolve transjurisdictional problems across borders and exchanging information on best-practice programs to resolve common urban problems can both improve the efficiency of local government and increase the competitiveness of private firms in the region. Several programs are worth presenting here.[10]

European cooperation networks. This program, introduced by the European Community (EC) in 1991, provides supplemental financial support to some 37 networks to promote the economic performance of less-favored regions by transferring technology and expertise. The areas covered include economic development, planning, transportation, environment, tourism, education and training, emergency relief services, and so forth. The networks are based on a wide variety of themes, ranging from Demilitarized and Environet to Euroisles and Universities-Regions.

One of these networks that has attracted attention in North America is Eurocities. Established in 1986 as a forum for European "second cities," it now has a membership of more than 40 large cities throughout Europe. The Eurocities network, which has received cofunding from the EC for certain subprojects, carries out three types of activities:

1. Experience transfer (e.g., best practice in the area of economic promotion)
2. Lobbying in the EC context (e.g., for the development of an effective EC urban policy)
3. Project development in the areas of transportation and communications infrastructure involving more than one city (e.g., the development of a high-speed rail line between Lyon and Turin)

Apparently, the Eurocities network has been successful; its membership has become so large that policy formulation and implementation have become increasingly difficult. In response, another cities network, Eurometropoles, has emerged with a more focused mission and a smaller membership.[11]

Urban pilot projects. Although the needs of cities seem to be unlimited, EC resources are not. Thus, because it cannot and should not do everything—in agreement with the EC's principle of subsidiarity (i.e., each higher level of government should do only what lower levels cannot do for themselves)—the EC recently began to fund pilot projects to test concepts designed to improve the effectiveness of urban policy in three problem areas:

1. Lack of access to jobs and training by people who live in peripheral and inner city neighborhoods
2. Increasing economic prosperity while preserving the environment
3. Revitalizing economic life in historic city centers

The results of the projects will then be disseminated to cities and member country governments and used in the formulation of local, national, and EC policies.

Cross-border cooperation. European border regions vary a great deal with respect to population density and economic development, but all tend to suffer from certain handicaps, including lower incomes and higher unemployment rates in their own national context; a peripheral position with respect to national economic and political decision making; and the problems imposed by the propinquity of different legal and administrative systems, poor cross-border communications, lack of coordination in public services, and differences in culture and language.

In the context of the EC's push toward a single market, borders are losing much of their historical significance, but the transition to a borderless economy will take some time. Thus, in 1990, the EC launched a special border program, INTERREG, to promote cross-border cooperation for economic development.

It is important to note that the EC, as a common market, has both internal borders (i.e., borders between member countries, such as France and Germany) and external borders (i.e., borders between member and nonmember countries, such as Germany and Poland). The economic differences between member countries are much less than between member and nonmember countries.

The creation of the single market is expected to improve the position of regions on internal borders but may create unfavorable conditions for those on the external borders as they move from the periphery of a national market to the periphery of the much larger EC market. In addition, the borders of the former Soviet bloc countries were essentially sealed until just a few years ago and cross-border cooperation was virtually nonexistent. Nevertheless, the INTERREG program is designed to help both types of border regions to better confront the opportunities and challenges presented by increasing economic integration.[12]

One of the most complex border regions in Europe is situated at the intersection of Switzerland, France, and Germany. The *Regio,* which was founded on the Swiss side in 1963, is based on strong cultural ties including a common history, language, literature, folklore, and architecture. After 30 years of collaborative activity, projects in economic development, transportation infrastructure, environment and energy, culture, and media/communication now involve the governments at all levels, private firms, and universities (Regio Basiliensis, 1988).

City Strategies

Every city-region wants economic prosperity and a high quality of life. The problem is that both objectives are becoming more and more elusive. As noted previously, it is no longer acceptable for national governments to simply prime the pump with deficit spending to solve a nation's economic ills. Increasingly, cities must depend on their own resources for solutions.

Most people in developed countries now live and work in cities. Thus, if a nation is to be competitive, its cities must be competitive. Similarly, if a city is livable and competitive, it is because local decision makers, working together and with others outside the region, make it happen.

The question is how does a city in decline pull itself up by the bootstraps in this new environment? Simple answers to this question abound: Create a friendly business environment, provide incentives to attract new firms and retain existing ones, expand exports, and internationalize.

Unfortunately, none of these alone is likely to provide a large, economically diversified city with adequate levels of prosperity and quality of life (QOL). In fact, some of these proposed solutions could raise one and lower the other or, in the long term, reduce both. Most cities these days are after what is increasingly referred to as *sustainable development,* although this concept has not yet been made operational at the local level.[13] Nevertheless, for most cities in the developed world, a high QOL—a "city that works" for both business and the people who live there—has become a necessary and perhaps even sufficient condition for achieving economic prosperity.

The response of many cities in this situation has been to initiate a process leading to a city strategy. Usually, four steps are involved: analysis, formulation, implementation, and evaluation—all of which draw on the specialized talents of individuals and institutions in the region and can be openly discussed so that the process is transparent and the strategy itself receives wide public acceptance.[14] This process, using the specialized knowledge of individuals from business, labor, government, and academia, can be organized in a variety of ways but must be approached as a long-term, multiyear project, leading to fundamental changes in public policy.

Phase #1: Analysis. In this context, *analysis* means research carried out in collaboration with a specially constituted task force drawn from all sectors of the economy. The objective of the research is to determine, on the basis of the city's actual and potentially available resources, the city's basic function(s) in the global, national, and regional economies as well as to identify those industry sectors—and clusters of industries—that are likely to be the strongest in the medium and long term. In other words, what are the city's competitive advantages now, and what could they be in the future with public support?

In this regard, Kresl's (1991) analysis of the concept of the gateway city (i.e., location on a physical or artificial border) is useful, especially the distinction he draws between the following:

- "Bridge cities" such as Buffalo and Seville: "cities which serve as conduits between two economies" (p. 351). These require significant investments in physical infrastructure such as ports, highways, and bridges.
- "Points of access cities" such as Montreal and Copenhagen: Cities that serve "as a point from which economic actors may gain access to cities in the other country." These require "office space, housing for high income individuals, urban cultural amenities such as museums, galleries, a concert hall, recreation facilities and parks and a good educational system" to attract business and financial services companies (p. 352).

Two views exist regarding the types of firms that a city-region can expect to attract and retain. The first one maintains that "the competitive position of regions is largely determined by comparative cost advantages, technological progress, agglomeration economies and the creation of a specialized market niche" (Nijkamp, 1990, p. 4). Traditionally, this has meant that firms, especially high-tech firms, locate in those regions in which costs are reasonable and other conditions are appropriate to creating an innovative milieu vis-à-vis specific product lines (e.g., biomedical or biotech). The other view, however, assigns more importance to QOL factors:

> According to a new (EC) study, firms are influenced by a wide range of location factors in deciding where to invest, and the factors which are of most importance vary considerably between projects. In general, location factors fall into three groups: market-related factors (access to EC market, strong national market, proximity to major customers and suppliers, general economic climate), the transport and communications infrastructure and labor. When it comes to the final choice of location, qualitative factors appear to be more important than quantitative (cost related) factors, firms being prepared to trade off cost disadvantages to secure qualitative advantages such as a pleasant climate and a good general living environment. (Commission of the European Communities, Directorate-General for Regional Policies, 1991, p. 55)

The analysis should include a map of the city's economic space (i.e., determining the geographical locations of the region's most important trade, financial, and technological partners) as well as identification of the region's main actual and potential industrial clusters. This information regarding existing relationships, together with information on expanding national and international markets, can be used to identify potentially competitive activities and redirect public policy to foster emerging industry sectors.

Phase #2: Formulation of a plan. At this point, the economic analysis must be synthesized and brought together to form a consistent and comprehensive plan for future actions. The basic question here is this: What can public policy do to strengthen the current and potential competitive advantages of the region? Other questions to be answered include the following:

- Is additional infrastructure needed to attract new industries or retain existing ones? If so, how much will it cost and how will it be financed? If there are competing uses for infrastructure dollars, which uses will yield the highest returns in incomes, jobs, and taxes for the region?
- Does the region have the necessary resources (i.e., water and energy) to support the emerging industry sectors in the future at reasonable costs? If not, what can be done?
- What regulations and laws need to be changed at the local or state level to increase the region's competitiveness without compromising the region's QOL?
- Given current trade patterns, what can be done to increase the region's exports and reduce its imports?
- What incentives are available to stimulate local firms to expand their operations locally and to attract new firms without compromising the region's fiscal base?
- Given the region's cost structure, positive and negative attributes, and needs to create both diversity and industrial clusters, what types of firms should be targeted for attraction to (or retention in) the local region?
- What educational-training policies are necessary to enhance the region's economy in the future?
- How can the region's universities and colleges be better used to support efforts to become more competitive nationally and internationally?
- What actions can be taken to increase the city's image and presence in international markets?

- What steps can be taken to ensure that the region's efforts to increase economic prosperity will be consistent with the need to achieve a high QOL?
- Is the city equipped to conduct its own national-international policy, with an office of protocol—an agency for managing sister city relationships and related matters?

It should be clear from this short list of questions that there are many trade-offs to confront. Given the diminishing resources available to the public sector, priorities must be established and hard decisions must be made. Isolated decision-making bodies, however, are frequently oblivious to or tend to ignore the negative consequences of their actions on the general planning process. Thus, it is important that some mechanism is created to impose consistency on the overall process. In this regard, collaboration between decision-making bodies is important, both horizontally (i.e., between groups dealing with different functional areas such as infrastructure and city marketing) and vertically (i.e., between various levels of government). Ideally, the region should strive for a private-public-academic partnership, developing positive synergies that pay off with enhanced prosperity and QOL.

Although each city's plan will be different, Table 6.5 provides a general outline of the main elements contained in a typical plan. The ideas in parentheses are meant to be suggestive, not exhaustive.

Phase #3: Implementation and evaluation. Depending on the depth and scope of the two previous steps, some activities can begin immediately, but full implementation may take 2 or 3 years. In fact, provision should be made to update information and reformulate the plan on a periodic basis so that future decision making is made on the basis of up-to-date information. Provision for objectively evaluating each component of the plan, using traditional and nontraditional indicators of economic prosperity and QOL, should be built into the plan from the outset.

It is clear that involving people who have an economic stake in the city's future can improve chances for the plan's success. This usually means including those enterprises that are the least mobile, such as labor unions and professional associations, public utilities (energy, transportation, and communications), locally owned media, and business services firms (law and accounting firms, banks, and freight forwarders).

The endeavor should include local universities that can provide specialized expertise on a wide range of topics from science and technology to urban planning and international business strategy. Advanced students

TABLE 6.5 Main Elements of a City Strategy

Overview of current competitive position and future prospects
- Strengths (city functions, strong sectors, location)
- Weaknesses (transportation facilities)
- Needs (infrastructure, complementary business services)
- General strategy (what will be done; who will do it; etc.)

Objectives (should be well defined)
- Economic (jobs created, firms retained or attracted)
- QOL (highway congestion, educational system)

Action components
- Legislative (improving the general business environment)
- Intergovernmental (developing vertical-horizontal synergies)
- Infrastructure
  - Physical (transportation, ports, communications)
  - Information (data sets, existing studies, accessibility)
  - Social (education and training, health care, social services)
- Technology-innovation (incubators, venture capital, networks)
- Export promotion (services to producers, expositions)
- City marketing (targeting of firms, investment incentives)
- Cultural (marketing ethnic diversity, internationalism)

External relations
- Protocol
- City image
- International events

Organization
- Participants-leadership
- Tasks-responsibility
- Timeline

Financing-budget

Evaluation and update of plan
- Economic prosperity indicators
- QOL

can serve as staff interns helping to organize the various committees and task forces that will be formed.[15]

Conclusions

It should not come as a surprise that those cities that most need new directions and processes are among those that are least capable of formulating and implementing them. Reasons for this inability to adapt are as follows:[16]

- The need to respond to short-term budget crises that overwhelm the system's ability to overhaul governmental structures and functions
- Out-of-date governmental structures themselves serving a variety of public and private vested interests, making wholesale restructuring in the short or medium term unthinkable
- Lack of experience, expertise, and belief in planning for regional economic development, including economic analysis, precluding interest by elected officials and bureaucrats who are ideologically committed to "the magic of the market" and opposed to virtually any type of planning
- An unfounded belief that simply the passage of time—riding out the business cycle—will restore prosperity to the region as it has many times in the past

Nevertheless, many cities are working to overcome these and other obstacles. The main ingredients in responding successfully to the current situation may be those that are the most difficult to measure or define: leadership and the development of positive synergies, which in turn lead to the efficient mobilization of the region's resources to produce a city-region that works.

NOTES

1. The major changes are well known: First and foremost is the long-term tendency toward increased internationalization of the economy throughout the post-World War II period, which was fomented by the United States and by the two main multilateral economic institutions that came into existence after World War II: GATT and the International Monetary Fund (IMF). This tendency was reenforced since the 1980s by the formation and/or deepening of regional trade blocs in both Europe and North America. Second is the dissolution of the Soviet Union in the early 1990s, which has had the most effect on Europe but, of course, has affected the entire world. Third, the emergence of supply side economics in the United States and Mexico's debt crisis in the early 1980s not only influenced North American economic policies but also provided added impetus to market-oriented, liberalization, and privatization policies throughout the world.

2. Throughout this chapter, the terms *city* and *region* will be used interchangeably. Each region, however defined, has a leading city and a hinterland as well as smaller cities. The identity of the region, however, is usually defined by the use of the leading city's name (e.g., the city of San Diego in the San Diego region).

3. The dramatic decline of (real) oil prices since the early 1980s has probably accelerated many of these trends as well.

4. Economic or industrial clusters can be defined as geographical "concentrations of competing, complementary and interdependent firms across several industries, including suppliers, service providers and final product manufacturers" (Morfessis, 1994, p. 33).

5. Clearly, the growing interdependency of capital markets limits national autonomy in this sense as changes in interest rates frequently result in changes in exchange rates—if they are flexible—as short-term capital moves to take advantage of the differentials.

6. The high (more than 10%) rates of unemployment in most European countries are often blamed on overgenerous welfare benefits, overregulation, and inefficient state-owned industries; there is, however, considerable opposition to dismantling these policies. In the United States, antistatist sentiment is also high; nevertheless, the Clinton administration openly supports the expansion of certain policies (e.g., health care and training programs) that could increase the government's role in certain sectors of the economy.

7. In 1985, an international group of scholars was founded in Paris called GREMI (*Groupe de recherche européen sur les milieux innovateurs*), which is engaged in the study of technological innovation processes and policies at the regional and local level. For an example of recent GREMI publications in English, see Aydalot and Keedle (1988).

8. One study estimates the cost of incentive packages provided by local governments vying for new automobile plants in midwestern U.S. cities during the period 1978 to 1990 at $3,900 to $108,000 *per job* (Glickman & Woodward, 1989).

9. Urban networking is not unknown in the United States or Canada; both countries have associations that exchange information and lobby on behalf of urban interests. Nevertheless, both the scale of the effort and the emphasis on the internationalization of cities in Europe seem to far outweigh anything that now exists in North America. An organization called North America Cities International was proposed, however, at a founders' conference at Wingspread Conference Center, in April 1992, by a group involved in the New International Cities Era project.

10. Information packets on each of these programs is available from the Directorate-General for Regional Policies (DG XVI) in the Commission of the European Communities in Brussels. The program summaries presented here were taken from that information and an interview with Dr. Marios Camhis, head of division, in July 1993.

11. Eurometropoles was established in 1990 with a membership of about 20 cities focusing on promoting economic development through agglomeration policies involving closer relations between local governments, private firms, and universities. Much of the information presented here on Eurocities and Eurometropoles was provided in an interview with Pierre Yves Tess of the Chamber of Commerce and Industry of Lyon in July 1993.

12. A recent report (Martinos & Caspari, 1990) summarizes the results of an EC-commissioned, two-part study on cross-border cooperation in Europe. Part I reports on the "characteristics, problems and initiatives in border regions" whereas Part II "develops a strategic approach to cross-border cooperation and makes a number of suggestions concerning the types of action and institutional arrangements" (p. 2). This 84-page report provides an excellent basis for analyzing border situations and developing cooperative strategies useful for academics, practitioners, and decision makers in border regions throughout the world.

13. See Calavita's (1993) discussion of this issue with respect to San Diego and other U.S. cities in his chapter "Measuring 'Quality of Life' in San Diego." My own operational definition of sustainable development is this: that level and composition of current output that is consistent with preserving the environment for the benefit of future generations.

14. This section draws heavily on Kresl (1992). Kresl's book can be helpful to the reader interested in more detail on most of the subjects covered in this chapter. Kresl, however,

refers to the four-step process outlined here as strategies for city internationalization. Although greater internationalization must occur in most cities of any size, empirical analysis might reveal that San Diego's best actual and potential customers are in adjoining regions (e.g., Los Angeles and Tijuana).

15. Clearly, not all universities, nor all faculty, are open to this type of partnership; budget cuts and downsizing, however, have affected academia as well as business and government, forcing structural changes.

16. These comments draw on my perusal of the relevant literature and firsthand observations of the San Diego-Tijuana situation, which is documented and analyzed elsewhere in this volume. See also Clement and Zepeda (1993).

REFERENCES

Aydalot, P., & Keedle, D. (Eds.). (1988). *High technology industry and innovative environments.* London: Routledge.

Calavita, N. (1993). Measuring "quality of life" in San Diego. In N. C. Clement & E. Zepeda (Eds.), *San Diego-Tijuana in transition: A regional analysis* (pp. 17-30). San Diego, CA: San Diego State University, Institute for Regional Studies of the Californias.

Clement, N. C., & Zepeda, E. (Eds.). (1993). *San Diego-Tijuana in transition: A regional analysis.* San Diego, CA: San Diego State University, Institute for Regional Studies of the Californias.

Commission of the European Communities, Directorate-General for Regional Policy. (1991). *Europe 2000: Outlook for the development of the community's territory.* Brussels, Belgium: Author.

Glickman, N., & Woodward, D. (1989). *The new competitors.* New York: Basic Books.

Kresl, P. K. (1991). Gateway cities: A comparison of North America with the European Community. *Ekistics, 58*(350-351), 351-356.

Kresl, P. K. (1992). *The urban economy and regional trade liberalization.* New York: Praeger.

Martinos, H., & Caspari, A. (1990). *Cooperation between border regions for local and regional development.* Brussels, Belgium: Commission of the European Communities, Directorate-General XVI.

Morfessis, I. (1994). Cluster analytic approach to identifying and developing state target industries: The case of Arizona. *Economic Development Review, 12*(2), 33-37.

Nijkamp, P. (1990). Spatial developments in the United States of Europe: Glorious victories or ignominious defeats? *Papers of the Regional Science Association, 69,* 1-10.

Regio Basiliensis. (1988). *Regio report: 1963-2013.* Basel, Switzerland: Author.

Reich, R. (1990). *The work of nations.* New York: Vintage.

Urban networking in Europe I: Concepts, intentions and new realities [Special issue]. (1991). *Ekistics, 58*(350-351).

Urban networking in Europe II: Recent initiatives as an input to future policies [Special issue]. (1992). *Ekistics, 59*(352-353).

7 World Cities in the Making: The European Context

ARIE SHACHAR

Introduction: The World City Concept

The world city concept is a recent revival of an old term that was well known in the European literature (Hall, 1966; King, 1990). The term was used to describe cities of distinct cultural-religious significance, such as Rome and Paris, or major capital cities that controlled and dominated imperial economies, such as London, Madrid, and Vienna. The resurgence of interest in this concept stems directly from Friedmann's research and publications in the 1980s (Friedmann, 1986; Friedmann & Wolff, 1982). Friedmann's main contribution was in relating a particular type of urban development to the global economic restructuring process. The world city was identified as the territorial expression of the emerging global economy. The main role of these cities was to cope efficiently with the complex management needs of the new economic system of global reach (Sassen, 1991).

The present complexity of the international economy is a direct outcome of the new international division of labor, the emergence of new industrial production sites in many of the less developed economies, and, thereafter, the establishment of new international trade patterns. The new international division of labor is spurred by the investments and activities of multinational corporations that spread out from the developed world to reach remote quarters of the less developed countries (Dicken, 1994). The spatial dispersal of industrial activities carried out on a global scale, the complex management needs of the big multinationals, and the emergence of a truly global market in financial transactions—all these economic restructuring processes caused the articulation of a new command and control structure, expressed in a new world urban hierarchy (Dicken & Thrift, 1992). In their new conception, the world cities constitute the spatial concentration of the various global control activities—the headquarters of national and transnational corporations and of multinations, the stock exchanges and the major financial institutions, the data processing and communications centers, and the full range of support facilities, mainly of the advanced producer services such as legal services, accounting, real estate companies, advertising, and marketing. World cities can thus be identified as the control, command, and management centers of the global economy (Smidt, 1992).

The European Context

The concept of world cities, which resurfaced in the United States in the 1980s, began to be applied extensively to the study of individual cities. These either were defined a priori as world cities, such as the global triad of New York, Tokyo, and London (Fainstein, Gordon, & Harloe, 1992; Mollenkopf & Castells, 1991; Sassen, 1991), or were the outcome of an extensive typological analysis, such as *Les Villes Européennes* (RECLUS/DATAR, 1989) and the Institute for Regional Planning (IRPUD; University of Dortmund, Germany) study (Kuntzmann & Wegener, 1991).

The well-publicized *Les Villes Européennes* study was a pioneering attempt to propose a full typology of European cities on the basis of a large number of urban characteristics. Some of these were related to economic activities, such as the number of persons employed in management, engineering, consulting, and research and development activities; the number and size of high-tech industrial parks; and the volume of trading in equities and bonds in the various financial centers, and so forth. Other characteristics were in the realm of infrastructure availability, such as the volume of airport traffic, the trade fairs and exhibitions, and the telecommunications facilities. The study also included cultural data on universities, press and publishing activities, and a composite index of cultural influence—festivals, museums, theaters, and musical performances.

Without going into a detailed critique of the ranking methodology, which suffers from some inherent weaknesses, it is still worthwhile to mention the ranking of the European cities because of the enormous popularity of this study and the widespread publicity it received since its publication. In the general classification, based on the summation of all the 16 variables, the rank-order is as follows: The uppermost echelon of the European cities includes only two cities: London and Paris. The second level is composed of the following cities: Milan, Madrid, Munich, Frankfurt,

Rome, Brussels, Barcelona, and Amsterdam. It emerges clearly from this study that in the 1980s, Europe had two world cities—London and Paris—and a second tier of international cities, spread throughout Western Europe, representing either national capitals or cities with a strong international orientation. The finding that the two cities on the uppermost level of the European urban hierarchy are capital cities, as also are four of the eight cities in the second tier, hints to the strong relationship prevailing in Europe between a world city status and strong national intervention in enhancing this status.

A much more sophisticated study of the European urban typology was carried out by an Italian team as part of a research project on cities and economic development (Conti & Spriano, 1990). Analyzing technological innovations as a multidimensional phenomenon, Conti and Spriano identified a large number of variables that through a factor analysis, allowed the identification of a new European urban hierarchy. The analysis was based on economic measures, such as the number of headquarters of the largest 500 industrial companies, of the 300 largest commercial companies, of the top 100 insurance companies, of the top 100 transportation companies, and of the top 100 advertising companies. Social measures were also applied, such as unemployment rates, net migration balance, annual rates of population change, housing obsolescence, proportion of rented housing, and number of cars per 1000 inhabitants. The first factor that came out of the statistical analysis, accounting for more than 40% of the variance, was most suitable in establishing the European world cities hierarchy because it was based on the two most important characteristics of world cities, the control functions and the advanced producer services functions. The calculation for each city of the scores for the first factor enables a ranking of the European cities along the axis of world city characteristics: Topping this rank-order with the highest scores are London and Paris. On the second level was a cluster of cities that included Brussels, Amsterdam, Rome, Frankfurt, and Milan. Following the Conti-Spriano study, it is possible to define London and Paris as world cities and Brussels, Amsterdam, Rome, Frankfurt, and Milan as cities with high scores in the world cities characteristics' scale.

In the last few years, several more attempts were made to rank European cities according to a combination of various urban characteristics. A few of these studies should be mentioned: *Urban Problems in Western Europe* (Cheshire & Hay, 1989); *City Audit* (Lever, McGregor, & Paddison, 1990); "West European Capital" (Palomaki, 1991); *Urbanization and the Functions of Cities in the European Community* (Parkinson, Bianchini,

Dawson, Evans, & Harding, 1992); and a recent survey of the best cities in the world in which to locate international businesses ("The World's Best Cities for Business," 1994). Most of these studies applied a plethora of economic, demographic, and social variables to determine the status of cities regarding their economic performance, social well-being, attractiveness to international business, and quality of life. With the growing awareness of the importance of the globalization of the economy to urban development, most of these studies include specific indicators to measure the level of the internationalization of a particular city. These indicators include, for example, the existence and the number of headquarters of multinationals, of international banks, of financial institutions, and so forth, as well as the level of accessibility of the city by international flights and, more recently, by superfast trains.

Reviewing the various attempts to establish the present structure of the European urban system and to build European league tables for cities provides a wealth of information on many facets of the urban development in Europe. On the other hand, some of the typological studies lacked a conceptual framework for the ranking operation and applied soft techniques in combining the variables into a meaningful composite index. Nevertheless, as most of these studies used variables that reflect the level of internationalization of the urban economy, it is possible to accept the empirical findings and to extract from the various studies a short list of the European cities having the strongest international orientation. This list includes London, Paris, Frankfurt, Randstad-Holland (the agglomeration of Amsterdam, Rotterdam, The Hague, and Utrecht), Milan, and Brussels. These cities might be termed *European world cities* or *European world cities in the making.*

Although it is not of major importance that a certain city is accorded the status of world city, it is meaningful to identify its new economic base of management and financial and business services activities. This particular transformation of the urban economy and, consequently, of the urban society and culture, links the local economy strongly with the global one, making these cities the territorial expression of globalization processes.

The concept of world city offered by academic circles was readily adopted by politicians and planners for use not only as an analytical tool but also as a policy-generating one. Maintaining the status of world city, or enhancing a particular city to move toward that status, has become an explicit metagoal of the highest importance in national and urban planning. In Europe, contrary to the United States, national intervention is well accepted to strengthen the competitive edge of a specific city within the global competition for control, management, and financial activities. Therefore, it is hardly surprising that the two European cities that are defined by all as world cities—London and Paris—have recently initiated well-articulated strategies to solidify their international and global role. In London, the formulation of this strategy was made public in the book *London: World City Moving Into the 21st Century* (HMSO, 1991), whereas in Paris the new plan for the Paris region, *Livre Blanc de l'Ile de France* (DRIEF/APUR/IAURIF, 1990), focuses on policies to ensure its international role. It is possible to place within the same policy context the proposal of the Dutch National Planning Authority to designate the three major cities of the Netherlands—Amsterdam, Rotterdam, and The Hague—as "international centers" (Ministry of Housing, Physical Planning and Environment, 1988).

World Cities Characteristics

Following the identification of the major characteristics of world cities, all related to restructuring processes of the economy and reflecting the territorial expression of globalization processes, it is possible to propose an operational definition of a world city as a large urban agglomeration specializing in international control capabilities, expressed in the urban milieu by three strongly interrelated components: a management and financial center of global reach, an extremely high concentration of producer services, and a rich physical and social infrastructure. It should be emphasized that almost all large metropolitan areas exhibited some elements of the three defining characteristics, but only the few cities that had all these components developed to the highest level can be defined as world cities. The other cities of an urban system can be placed along the continuum of world city characteristics, according to their particular score on this continuum. The concept of world city is thus moving from a dichotomous one to a continuum of world city characteristics, along which each city scores according to its own defining features.

World cities, being the territorial expression of globalization processes of both an economic and cultural nature (Robertson, 1992), can be characterized as composing the following features:

1. A spatial concentration of international management activities: The growth and reorganization of corporations into divisions and the globalization of markets entail a complex organizational structure, its managerial

responsibility extending over a large territorial reach. The more complex the organizational structure is, the wider is its territorial extent and the higher is the demand for a specialized management function that is totally dependent on the availability of a huge amount of up-to-date information in managing the corporation (Morton, 1991). The headquarters of major corporations tend to agglomerate in information-rich environments that are the business centers of the world cities (Castells, 1989; Mulgan, 1991). The functional hierarchy of corporate organizations is becoming a cornerstone for defining cities along the continuum of world city characteristics. A world city of the highest order will be defined by the spatial agglomeration of the management centers of the largest and the most internationally oriented corporations and multinationals.

2. A spatial concentration of financial institutions: Financial activities remain the most mobile of all economic activities. Recent breakthroughs in telecommunications and the development of trading systems that are carried out on-screen without a trading floor seem to allow the dismantling of the actual physical places into which financial activities are tied (O'Brien, 1992). The NASDAQ system in New York is an illuminating example of the new possibilities of on-screen off-floor trading. Nevertheless, financial institutions still retain their spatial proximity and tend to concentrate in a limited area, a direct outcome of the unavoidable need for face-to-face exchange of information and the intense personal communication that is at the base of the mutual trust among traders and dealers (Budd & Whimster, 1992). The territorialization of the financial activities, elaborated on later in this chapter, is one of the most important roots for the evolution of the world cities. It is sufficient to state at this point that the spatial juxtaposition of the international management centers and the international financial centers create the new Marshallian districts that are at the core and essence of world cities (Amin & Thrift, 1992).

3. A spatial concentration of advanced producer services: Big corporations, dealing with complex production and distribution activities in many parts of the world, increasingly tend to externalize most of their professional support activities. The more internationalized their business becomes, the greater is the need for highly specialized services in the legal, accounting, technical, and consulting sectors. This demand stimulates the growth of the advanced producer services sector (Daniels, 1991). The high level of specialization of this sector allows a significant reduction in the transaction costs of the management and control activities and, consequently, intensifies the interaction between the headquarters of corporations and the highly specialized producer services. This boosts the producer services

into closer proximity to and intermingling with the management and financial centers. The combination of management centers, financial centers, and producer services' offices constitutes the economic and physical core of the world cities.

4. Physical and social infrastructure: A world city is characterized by the rich supply of infrastructure and of sophisticated urban amenities. One critical aspect of the physical infrastructure is a top location for offices, which can be described as a defined area within the central business district with a high concentration of high-rise, modern office buildings of architectural distinction and prominence. The spatial concentration of distinctive office buildings dominates the urban skyline and conveys a strong image of power and prestige (Knox, 1991). This extraordinary concentration of modern office buildings is aimed at creating a high-quality business environment and providing the critical mass of activities for personal interaction and shared value of prestige. The attraction of a top location for business is based on the combination of several closely related elements: high prestige, pleasant and impressive ambiance, and the best facilities for employees and visitors. Developing a new top location, such as La Défence in Paris, Cannery Wharf in London (Daniels & Bobe, 1993), and Finanzplatz in Frankfurt, requires huge investments that can be mobilized only through international sources, thus strongly linking the international financial centers and the real estate markets (Coakley, 1994). The magnitude of the investments in developing a top location creates a major stimulus in the urban economy and provides a large number of employment opportunities. The investments in building a top location can come from private or public sources, or from a combination of both, but they always require a long-term public commitment to provide for the transportation infrastructure and facilities to serve the new top location.

World cities are characterized not only by major elements of physical infrastructure but also by a rich assortment of social and cultural infrastructure. This element of the urban milieu caters to the demand and aspirations of the local middle- and high-income groups, which are growing steadily in the wake of the professionalization of the service industry, as well as to the demand of the international elite groups of directors, managers, and experts who move around between the international management and financial centers. The intensive networking of the local and international professionals is enhanced by the availability of a large number of meeting places in the form of clubs, restaurants, bars, coffee houses, sport facilities, and so forth, in which the informal contacts and the social network can be maintained and pursued. As for the cultural infrastructure, world cities are richly endowed with the largest variety of cultural and entertainment facilities of the highest quality, such as museums, galleries, opera houses, theaters, and concert halls. This spatial concentration of cultural and entertainment facilities, capable of catering to a large number of visitors of the most diverse tastes, allows world cities to become major attractions for urban tourists from all parts of the world. Urban tourism, becoming a major facet of the global economy, thus enhances the international role of world cities and significantly strengthens their urban economic base. The growing importance of urban tourism in shaping a world city status will be discussed further later in this chapter.

European Financial Centers

The theoretical elaboration of the characteristics of world cities identified the financial and management sectors of the urban economy as the main indicators of the position of a city on the world city scale. The major activities carried out in a financial center, such as the mobilization of capital resources, the trading in money, and the securement of investments through future contracts, are the main stimuli for the efficient operation of national economies and for the global economy at large. Therefore, all the developed countries as well as some of the newly industrialized countries encourage the establishment and growth of a national financial center. Those financial centers that have become internationally important because of their size, volume of transactions, multitude of their financial instruments, level of innovation, and the global reach of their activities and clients are propelling the urban center in which they operate toward a world city status.

When dealing with European world cities, it is most productive to identify the major international financial centers and to follow the dynamics and the competition among these centers (Lee & Schmidt-Marwede, 1993). Financial businesses tend to be highly concentrated on a national scale. This spatial concentration is even more pronounced at the global scale because the three global centers—New York, Tokyo, and London—dominate to a large extent the entire global scene.

The persistence and growing concentration of financial services in a small number of centers indicate the existence of strong agglomeration pressures of this sector. These tendencies have been studied theoretically and empirically in the last generation, culminating in the *City Research* project (see London Business School, 1992). The benefits of concentration are related to external economies of financial activities. The most important externalities, which are most often mentioned in locational surveys, are access to an extensive labor pool, access to supporting services, strong personal contacts, and some "prestige of location."

Access to labor. Financial services tap large pools of highly skilled labor as well as a large number of office staff to carry out routine and semiskilled work. The extent of this labor pool is exemplified in London, where in 1990, about 800,000 people were employed in banking, finance, insurance, and business services. This extremely high number of employees in financial services accounted for about 23% of the city's total employment. In Paris, in the same year, the size of the labor pool in financial services was about 190,000; the corresponding figure for Frankfurt was about 35,000 (HMSO, 1991, chap. 2).

The large size of the financial services labor market allows corporations and firms a high level of flexibility in establishing teams with highly specialized ad hoc tasks to perform, which is the cornerstone of any major financial center's activities. The reliance of the financial sector on a large labor pool, part of it highly skilled and specialized, means that world cities must have a significant population size to allow the recruitment of the large number of employees required in the financial sector.

Access to supporting services. The tendency of financial firms to cluster arises from the approach to externalize their most specialized activities, such as legal advice, international accounting, information services, specialized printing, marketing and advertising, computing and programming consultants, and other technical services. These highly specialized advanced producer services have to be located in proximity to the various financial firms because of the strong and frequent interaction between them. The headquarters of the management centers of big corporations, many of them concentrated in the central cities of the large metropolitan areas, create a strong and persistent demand for advanced producer services. The concentration of advanced producer services thus constitutes a major component of European world cities (Moulaert & Todtling, 1995).

Personal contacts. Financial transactions require the acquisition and use of huge amounts of information, most of which is channeled through formal links and information networks. These information flows are open and available to everyone in the financial district and, thus, do not provide a competitive edge among the large number of competing firms. Another source of information, which flows through informal personal contacts in the workplace and in restaurants, bars, clubs, shops, and other meeting places for professionals, could be highly valuable in obtaining information still unknown to the competitors. It can be argued that the financial district thrives because of its role as a locus of personal interaction (Amin & Thrift, 1992). The need for strong personal contacts is even more marked when new financial products are introduced in the market. In this respect, the financial district behaves as a locality in which personal interactions are based on mutual trust and understanding and on a common cultural and social orientation. The inherent need for information exchange through personal contacts might be the most important cohesion factor in maintaining the financial district as a compact spatial unit within the urban region.

Prestigious locations. In many of the surveys conducted among the decision makers of big corporations and of financial institutions, reference is made to the importance of a prestigious site in their locational decisions. The prestige of a site is related to its proximity to the anchors of the management and financial community (a central bank, the stock exchange, etc.), as well as to the high quality of urban design and to the outstanding architecture of a top location environment.

Some of the major European cities made a deliberate effort in the last few years to build and maintain new office complexes of the highest architectural quality to attract headquarters of multinational corporations and the most influential financial institutions to their cities. These efforts demanded huge public and private capital investments, both in the buildings themselves and in expanding infrastructure systems of transportation and telecommunications. The two cities at the top of the hierarchy of European world cities—London and Paris—provide striking examples of the attempt to develop a top location of high architectural prestige. Cannery Wharf, in the Docklands area of London, and La Défence, west of Paris, stand out as the shining realization of the competition among the world cities to attract corporations of global reach. On a smaller scale, the Finanzplatz in Frankfurt fulfills a similar role. In the context of the "landscapes of power" (Zukin, 1992) of the European cities, these new glittering office complexes symbolize, more than any other urban facet, the unequaled might of international management and finance in the crystallization of world cities within the European urban system.

The global economy is growing and changing continuously, and the balance of power among trade blocs and national economies is fast shifting. Within this most dynamic world system, however, the major financial centers tend to maintain quite a stable pattern. London, New York, and Tokyo have kept their supremacy as the global financial centers for several decades. Within this triad, however, London differs from New York and Tokyo because a substantial proportion of New York's and Tokyo's financial activities are domestic in origin, whereas London is predominantly an international center because most of its financial operations are not directly related to the British economy (Pryke, 1991). Within Europe, London is, of course, the most important financial center, with growing competition waged by Frankfurt and Paris, and, to a lesser degree, by Amsterdam and Zurich. The European pattern of financial centers, crucial in identifying the European world cities, has remained quite stable during the last decades. Following is a review of the processes that brought about this pattern's emergence and apparent consistency.

The European financial centers are fiercely competitive, each of them trying to increase its market share in the various financial markets. This competition is based on the improvement of the comparative advantages of one center over the others (Leyshon & Thrift, 1992). The competitive edge of a financial center is determined primarily by the profit margins of financial transactions, the availability of the fullest range of financial instruments, and the highest level of expertise in conducting complex financial transactions in investment banking and in the hedging of financial resources. Another major element affecting the competitiveness of financial centers is the intensity of government intervention into their operations. Because a strong financial base is critically important for a national economy, governments tend to intervene in and control the operation of the financial centers. Regulations of numerous types are enforced on these centers, such as the level of liquidity demanded of various financial institutions, the application of the principle of a level playing field, the concrete interpretation of transparency in trading, the limitation of membership in the stock exchange, the roles and responsibilities of intermediaries, the taxes levied on transactions, and the procedures of settlements.

The crucial importance of curtailing government intervention and of deregulating most of the financial centers' activities became highly evident during the last 10 years in the renewed ascendancy of London, reinforcing its role as a global financial center. This was a direct outcome of the completion of a deregulation process that culminated in the Big Bang of

1986 when, in a dramatic move, most of the traditional regulations were abolished, fixed commissions were scrapped, and restrictions on entry to the stock exchange were removed. The Big Bang ensured the return of London's financial center to global supremacy, on a par with the two other global centers of New York and Tokyo (Bank of England, 1989).

Within the context of the competition among the European financial centers, the far-reaching impact of the Big Bang on London's position as a global financial center did not escape the attention of European governments and regulatory agencies. Since the end of the 1980s, all of them went into a deregulation process whose intensity depended on the political and economic climate of each country. Paris is an excellent example of this new approach as the French government phased out, at the beginning of the 1990s, most of the restrictions and regulations controlling the Paris stock exchange. The strong deregulation program made the Paris financial center much more competitive, bringing it closer to London in efficiency of its operation, thus allowing for a significant increase in its market share within the European system of financial centers.

The effects of the British government's intervention in initiating and implementing the deregulation of the London Stock Exchange allowed the city to regain its position as a global financial center, thus strengthening the role of London as a world city not only within the European context but also within the global one. This provides a prime example of the interaction between national policies and international performance. In London, the financial sector was singled out as the main stimulus for enhancing its international position. In other European cities, different avenues were chosen in expanding their international activities.

Because of the importance of this topic, it is worthwhile to take a short detour from the main direction of this chapter to discuss the French government's intervention to enhance the international position of Paris. Paris provides an example of a different type of government intervention to achieve the same goal of maintaining and bolstering the world city status of the French capital. This is done by huge national investments in cultural and entertainment facilities (museums, opera house, theme parks, etc.), through improvements in the internal accessibility of the city, and by making Paris the center of a national and all-European system of high-speed trains (TGV). Attracting headquarters of multinationals to Paris was facilitated by offering the most modern office center having the highest architectural distinction in La Défence quarter. All these government policies and activities are focused on the provision of enabling infrastructure of a cultural and physical nature.

The examples of government intervention in enhancing the international roles of the two prime European cities highlight the common approach prevailing in Europe of central governments playing a major role in improving the competitive edge of their capital cities by supporting their international attractiveness and business performance. Although this intervention was carried out in London by streamlining the market mechanism and boosting the major international sector of activities, it occurred in Paris, however, through direct investments in the city's infrastructure, aiming at improving the quality of life and level of attractiveness for international businesspersons and tourists. In some other European cities, similar actions are being taken to raise the level of their international activities, thus moving them upward along the continuum of world city characteristics.

An international financial center is characterized by a high share of cross-border transactions in international equity business, foreign currency exchanges, and international futures and options contracts. Reviewing the market share of the various European financial centers illuminates the following rank-order: In 1990, London's market share of international banking was 18.4% of the world total; Paris's was 6.7%; Frankfurt's, 5.0%; Zurich's, 3.9%; and Amsterdam's, 2.7% (Lee & Schmidt-Marwede, 1993). It is obvious that at the beginning of the 1990s, London had a most significant lead over all other European financial centers. Thus, in financial activities, London was unrivaled in the top position of the European world cities hierarchy. The big gap between London and the other European competing centers makes the challenge to its supremacy almost an unattainable goal, but to keep this supremacy, London will have to pursue the introduction of new financial instruments, push new technologies to their limits, and apply the highest expertise available to carry out the complex financial transactions.

The internationalization of the financial sector is most evident in the cross-border trading in equities. For many years, the bulk of trading in domestic shares was carried out in the local stock exchanges. This geographical pattern changed dramatically when London introduced, in 1985, the SEAQ International system, by which foreign shares were quoted and traded on-screen. In a matter of 10 years, a significant number of European shares have moved from their home country to be traded at the London Stock Exchange, as SEAQ International now deals with more than 90% of Europe's cross-border equity trading turnover ("The 1994 Guide to World Equity Markets," 1994). Analyzing the cross-border trading of the competing financial centers at the beginning of the 1990s reveals a distinctive picture: Although 48% of the total turnover in equity trading in London is in foreign shares, the respective figure in Paris is only 4.5% of the total turnover and in Frankfurt, only 3.6%. The major European stock exchanges lost a decisive proportion of trading in their own home stocks for the most part because trading in London is much more profitable as a result of its greater liquidity, depth of market, and lower commissions. The London Stock Exchange, through SEAQ International, captured a sizeable proportion of the trade in foreign stocks: 26.7% of the French home market, 25.8% of the Swiss, 18.7% of the Spanish, 12.2% of the German, and 11.7% of the Italian home market. Topping this list are the Dutch stocks, with 50.3% of the home market traded in London. The European financial centers, mainly the important ones—Paris, Frankfurt, and Amsterdam—are in the process of restructuring and streamlining their stock exchange operations to compete with London and to bring home the trading in their national stocks (McKinsey, 1992).

The fiercest competition among the European major financial centers occurs in the relatively new realm of derivative markets. The new financial instruments, such as futures, options, and swaps, are for the most part risk-hedging instruments that emerged because of the dangers inherent in a highly volatile financial environment. The derivative markets were well established in the United States, in which Chicago hosted the largest future exchange. In Europe, the derivative market was established only as late as 1982, when London launched its future exchange market (LIFFE), followed by the Paris-based MATIF and the Frankfurt-based DTB.

The data available for the number of futures and option contracts traded in the various financial centers show that London is the leading center in this new lucrative market. In 1993, London handled about 110,000,000 future and option contracts, whereas Paris was second in line with about 72,000,000 contracts; Frankfurt trailed in third place with about 50,000,000 contracts. London's lead is still evident, but the gap between London and Paris and Frankfurt is now closing fast. London was particularly successful in launching trading in nondomestic future contracts, holding more than 80% of the market in the German Bund, Italian Bond, and Euro-Swiss franc contracts. International contracts accounted for more than 53% of LIFFE's total turnover in 1992.

Because derivatives are newly developed financial instruments, borrowed from the United States and transplanted to Europe only in the last few years, it is clear that incumbency cannot play a major role in the competition among the various European financial centers. London is still leading in the derivative markets, but its dominance is much less secure than in the international equities and foreign exchange markets and, in specific types of markets, such as traded options, Frankfurt has already taken over the lead. The intense competition among the European financial centers in the most modern sector of derivatives is becoming the clearest example of the spatial struggle for the territoriality of the European world cities. This is carried out by their attempts to increase their market share in this lucrative and profitable global financial instrument. This struggle is strongly supported by national governments and various regulatory agencies.

The spatial competition between the various European financial centers in the derivatives market is fueled by the application of the most advanced information technologies and by the various centers recruiting experts at managing the complicated mathematical models by which the derivative market is being run. The future of Europe's financial system will be determined by the outcome of the fierce competition in the derivative market.

European World Cities as Destinations of Urban Tourism

Urban tourism is the third leg in the triad of urban economic bases that together with financial and management activities determine a world city status. Urban tourism is not a new phenomenon in the European way of life. The "Grand Tour" of European sites is a well-known example of early urban tourism. But during the last few years, a surge in the volume of urban tourists and a much sharper focus in their destinations have occurred. Urban tourism of international origin is a major determinant in the internationalization process of European world cities and is a significant component of the postindustrial urban economy.

The rank-order of European cities, in their attraction for visitors, can be established on the basis of the results of a recent survey carried out on behalf of the VVV (the tourism agency of the city of Amsterdam; see KPMG, 1993). The number of arrivals of visitors, in descending order (in millions), is as follows: London, 14.7; Paris, 12.6; Munich, 3.2; Rome, 2.7; Vienna, 2.6; Berlin, 2.5; Milan, 2.1; Brussels, 2.0; Frankfurt, 1.9; and Barcelona, 1.8. The cities in this list are clearly divided into two groups, the first including London and Paris, the cities of global attraction, with more than 12 million visitors per year. The second group includes all the other cities in the list, whose number of visitors is in the range of 2 to 3 million. This two-level hierarchy, determined solely by the city's attraction for visitors, corresponds quite well to the hierarchy of the European world cities in which London and Paris are defined as the two cities with the highest level in the scale of world city characteristics (RECLUS/DATAR, 1989; Parkinson et al., 1992). The identification of London and Paris as constituting the upper level of world city characteristics as well as having the highest level of attraction for visitors gives a strong indication of the relationship between urban tourism and the formation of world cities.

Some of the European cities are increasingly oriented toward the global economy and thus are moving ahead in the scale of world city characteristics. This international orientation is strongest in the sectors of financial, management, and control activities. Becoming major financial or management centers makes these cities a prime destination for business travelers who need face-to-face contacts to carry out their transactions. Therefore, the higher the cities are in the scale of world city characteristics, the larger will be the number of business travelers they attract.

Urban visitors are increasingly characterized by the almost inseparable combination of tourists and businesspersons, the two groups visiting and intermingling in the same city. This combination is even more evident in some cases in which a business trip, such as participation in a conference, turns into tourism when the conference is over. The business travelers, mainly those who are staying overnight, contribute significantly to the urban economy by availing themselves of the lodging, restaurants, and cultural and entertainment facilities. The segmentation of visitors between tourist and business trips, mainly in the big cities, is therefore becoming blurred and not too meaningful. From available statistics, it can be deduced that tourists constitute between 40% and 50% of the total visitors to most European cities, including the two European world cities, London and Paris, with the remaining 50% to 60% being business travelers. For a few European cities, the breakdown between business and leisure visitors is available (see Van den Berg, Van der Borg, & Van der Meer, 1994). Beds/nights in hotels of business travelers as a percentage of total hotel occupancy is as follows: Antwerp, 65%; Copenhagen, 62%; Glasgow, 75%; and Lyon, 92%. Edinburgh is the only exception in this list because its share of business travelers is only 30%, an indication of this city's strong attraction for leisure and tourism.

The internationalization of the urban economy of many European cities is strongly affected by the explosive increase of business travel, which has two main streams: One is the regular business trip needed for business meetings, consultations, management contacts, marketing, and sales. Participation in mass gatherings, in the form of conferences, conventions, exhibitions, and trade fairs constitutes the other stream. The regular business travel is the outcome of the need for personal contact between businesspersons, mainly those in the middle and upper echelons of their organizations. The more complex the business decisions become, the stronger is the need for interpersonal exchange. It is increasingly evident that in major business decisions, the recent technological breakthroughs in telecommunications cannot substitute for personal interaction but instead render face-to-face meetings more meaningful and prevalent. This type of business travel shapes the strong flows of business travelers among the major European cities in which the financial institutions, the headquarters of transnational corporations, and the multitude of the most advanced producer services are located (Urry, 1990).

The business traveler contributes significantly to the urban economic base because his or her spending power is considerably higher than that of the regular leisure traveler. The business traveler stays overnight at hotels, enjoys the cultural and entertainment facilities, and might even spend some time between meetings in sightseeing and visiting the city's museums and historical sites. Many business travelers indulge in shopping for specialized items of local character. It has been estimated that a business traveler spends three times more money than the leisure traveler (Petersen & Belchambers, 1990). The combination of the increased number of business travelers and their high spending capacity makes business travel one of the main elements in strengthening the economic performance of the European cities characterized by a strong international orientation.

Traveling for participation in conferences represents the second facet of business travel of both national and international organizations. Across the world and particularly in North America and Europe, conferences, conventions, exhibitions, and trade fairs constitute a major growth sector of the urban economy (Smith, 1990). A great diversity exists in the conference market. Many conferences of corporations, associations, and academic institutions are quite small, attended by only a few hundred members, whereas the conventions of major organizations and political parties may assemble thousands or even tens of thousands participants. Because of the large number of visitors involved and the relatively long duration of the meetings, this type of business travel is regarded as of major economic benefit to the cities in which those meetings convene, resulting in cities competing strongly to be the venue of these conferences (Var, Cessario, & Mauser, 1985).

To host a major conference, a city must fulfill several demanding requirements: a high level of accessibility, especially in air travel and, more recently, within the Western European context, in high-speed train networks; modern, spacious, and well-equipped conference facilities; and a large stock of high-quality accommodations. An attractive urban image, which will please not only the delegates but their accompanying persons, is also a major asset in the competition of cities to be the venue of national and international conferences (Law, 1992).

Few cities can offer the whole package of these requirements, giving the two major European world cities—London and Paris—the leading edge in hosting international conferences. The list of European cities arranged by rank-order of the number of international conventions held there in the 1980s is as follows: Paris, London, Geneva, Brussels, Randstad-Holland, and Madrid. Paris and London are well ahead of the other cities listed (Shaw & Williams, 1994). As attendance in international conferences becomes an integral part of the business and professional way of life, world cities will increasingly depend on their role as hosts of these global meetings for their economic strength.

The role of world cities as an attraction for visitors from all over the world is reinforced by large investments, public and private, in building and maintaining cultural and entertainment facilities of the highest level and distinction (Williams & Shaw, 1991). Paris is the prime example of this urban policy. Investing in cultural and entertainment facilities, such as the numerous museums, the new opera house, the restoration of historical quarters, the striking architecture of major new public buildings, and the establishment of Euro-Disney, has strongly enhanced the level of attraction of the city. A similar policy is being pursued by other European cities, although on a much smaller scale: Frankfurt is developing its new museum row along the Main; Amsterdam is putting tremendous efforts in keeping its historical core alive and attractive; London is investing, mainly privately, in its theaters, music, and visual arts.

The policy of encouraging the investment in cultural and entertainment facilities fulfills a double role: the attraction of visitors on a global scale and, at the same time, catering to the local and international elites residing in the city. This is important in attempting to influence the international elites when they consider the location of their corporation's headquarters. Various surveys of the decision-making process of locating the new offices of a regional division, headquarters of multinationals, and branches of major financial firms indicate the importance given to the quality of urban life and the availability of cultural and entertainment facilities of the city being considered in the final locational decision. The convergence of the two lines of reasoning, attracting visitors and influencing the international elite to move to and work in a particular city, is making the urban policy of investing in cultural and entertainment facilities an effective tool in the competition for a world city status.

Conclusion

This chapter has applied the concept of world cities to the European urban context. For many centuries, the European urban system had been influenced by the division of the continent into many national units, each nation bolstering the development of its capital city. Consequently, the size, the roles, and the relative importance of the capital cities were an outcome of national power in political and economic terms. Therefore, the European world cities in the 19th and the beginning of the 20th centuries were the metropolises of the powerful European empires, their relative position on the world cities scale determined by the might and wealth of the respective empires. The present emergence of European world cities is determined primarily by the new processes of economic globalization that are shaping a new world urban hierarchy. Analyzing the economic base, the functions, and the roles of European cities within the global economy indicates that London and Paris have regained their former status as world cities, whereas Frankfurt, Randstad-Holland, Milan, and Brussels are world cities in the making. Although the development of world cities in Europe is stimulated by economic restructuring processes and by the globalization of the economy, European countries are still involved in enhancing and strengthening the status of their major cities and in pushing them forward to become full-fledged world cities. The analysis of world cities within the European context provides an illuminating example of urban development shaped by the combination of fierce global competition and strong social regulation as a result of state and municipal intervention.

REFERENCES

Amin, A., & Thrift, N. (1992). Neo-Marshallian nodes in global networks. *International Journal of Urban and Regional Research, 16,* 551-587.

Bank of England. (1989, November). London as an international financial centre. *Bank of England Quarterly* [Bulletin].

Budd, L., & Whimster, S. (Eds.). (1992). *Global finance and urban living.* London: Routledge.

Castells, M. (1989). *The informational city.* Oxford, UK: Blackwell.

Cheshire, P. (1990). Explaining the recent performance of the European Community's major urban regions. *Urban Studies, 27,* 311-333.

Cheshire, P., & Hay, D. (1989). *Urban problems in Western Europe: An economic analysis.* London: Unwin Hyman.

Coakley, J. (1994). The integration of property and financial markets. *Environment and Planning A, 26,* 697-714.

Conti, S., & Spriano, G. (1990). *Effeto citta.* Turin, Italy: Fondazione Agnelli.

Daniels, P. W. (1991). *Services and metropolitan development.* London: Routledge.

Daniels, P. W., & Bobe, J. M. (1993). Extending the boundary of the city of London: The development of Cannery Wharf. *Environment and Planning A, 25,* 539-552.

Dicken, P. (1994). Global-local tensions: Firms and states in the global space economy. *Economic Geography, 70,* 110-128.

Dicken, P., & Thrift, N. J. (1992). The organization of production and the production of organization. *Transactions of the Institute of British Geographers, 17,* 279-291.

DRIEF/APUR/IAURIF. (1990). *Livre blanc de l'Ile de France* [The white book of the Ile de France]. Paris: Author.

Fainstein, S., Gordon, I., & Harloe, M. (Eds.). (1992). *Divided cities: New York and London in comparative perspective.* Cambridge, MA: Blackwell.

Friedmann, J. (1986). The world city hypothesis. *Development and Change, 17,* 69-84.

Friedmann, J., & Wolff, G. (1982). World city formation. *International Journal of Urban and Regional Research, 6,* 306-344.

Hall, P. (1966). *The world cities.* London: Weidenfeld & Nicolson.

HMSO. (1991). *London: World city moving into the 21st century.* London: HMSO.

King, A. D. (1990). *Global cities: Post-imperialism and the internationalization of London.* London: Routledge.

Knox, P. (1991). The restless urban landscape: Economic and sociocultural change and the transformation of metropolitan Washington, D.C. *Annals of the Association of American Geographers, 81,* 181-209.

KPMG. (1993). *Overview: Comparative analysis of 34 European cities.* Amsterdam: Author.

Kuntzmann, K., & Wegener, N. (1991). *The pattern of urbanization in Western Europe, 1960-1990* (Vol. 18). Dortmund, Germany: University of Dortmund, Institute for Regional Planning.

Law, C. (1992). Urban tourism and its contribution to economic regeneration. *Urban Studies, 29,* 599-618.

Lee, R., & Schmidt-Marwede, V. (1993). Interurban competition? Financial centres and the geography of financial production. *International Journal of Urban Regional Research, 17,* 492-515.

Lever, W. F., McGregor, A., & Paddison, R. (1990). *City audit: Comparative European urban performance data.* Glasgow, UK: University of Glasgow, CURR/TERU.

Leyshon, A., & Thrift, N. J. (1992). Liberalization and consolidation: The single European market and the remaking of European financial capital. *Environment and Planning A, 24,* 49-84.

London Business School. (1992). *City research: Project interim report.* London: The Corporation of London.

McKinsey Company. (1992). *Charting a course: Towards a greatly improved international position for the Dutch security market.* Amsterdam: Amsterdam Stock Exchange.

Ministry of Housing, Physical Planning and Environment. (1988). *Fourth physical planning memorandum (VINO).* The Hague, The Netherlands: SDU.

Mollenkopf, J., & Castells, M., (Eds.). (1991). *Dual city: Restructuring New York.* New York: Russell Sage Foundation.

Morton, M. S. (1991). *The corporation of the 1990s: Information technology and organizational transformation.* New York: Oxford University Press.

Moulaert, F., & Todtling, F. (Eds.). (1995). The geography of advanced producer services in Europe. *Progress in planning, 43,* 89-274.

Mulgan, G. J. (1991). *Communications and control: Networks and the new economics of communication.* New York: Guilford.

The 1994 guide to world equity markets. (1994, June). *Euromoney* (Suppl.).

O'Brien, R. (1992). *Global financial integration: The end of geography.* London: Chatham House.

Palomaki, M. (1991). On the possible future West European capital. *Geojournal, 24,* 257-267.

Parkinson, M., Bianchini, F., Dawson, J., Evans, R., & Harding, A. (1992). *Urbanization and the functions of cities in the European Community.* Liverpool, UK: Liverpool John Moores University, European Institute of Urban Affairs.

Petersen, J., & Belchambers, K. (1990). Business travel: A boom market. In M. Quest (Ed.), *Howarth book of tourism.* London: Macmillan.

Pryke, M. (1991). An international city going global: Spatial change in the city of London. *Environment and Planning D, 9,* 197-222.

RECLUS/DATAR. (1989). *Les villes européennes* [The European cities]. Montpellier, France: Author.

Robertson, R. (1992). *Globalization: Social theory and culture.* London: Sage.

Sassen, S. (1991). *The global city: New York, London, Tokyo.* Princeton, NJ: Princeton University Press.

Shaw, G., & Williams, A. M. (1994). *Critical issues in tourism.* Oxford, UK: Blackwell.

Smidt, M. de. (1992). A world city paradox: Firms and urban fabric. In F. Dieleman & S. Musterd (Eds.), *The Randstad: A research and policy laboratory* (pp. 97-117). Dordrecht, The Netherlands: Kluwer.

Smith, G. V. (1990). The growth of conferences and incentives. In M. Quest (Ed.), *Howarth book of tourism.* London: Macmillan.

Urry, J. (1990). *The tourist gaze: Leisure and travel in contemporary societies.* Newbury Park, CA: Sage.

Van den Berg, L., Van der Borg, J., & Van der Meer, J. (1994). *Urban tourism.* Rotterdam, The Netherlands: Erasmus University, Euricur.

Var, T., Cessario, E., & Mauser, G. (1985). Convention tourism modelling. *Tourism Development, 6,* 194-200.

Williams. A. M., & Shaw, G. (1991). Tourism in urban and regional development: Western European experiences. In S. Hardy, T. Hart, & T. Shaw (Eds.), *The role of tourism in the urban and regional economy. London: Regional Studies Association.*

The world's best cities for business. (1994, November). *Fortune, 130,* 60-81.

Zukin, S. (1992). The city as a landscape of power: London and New York as global financial capitals. In L. Budd & S. Whimster (Eds.), *Global finance and urban living* (pp. 195-223). London: Routledge.

8 Determinants of the Growth and Decline of Cities in North America

PIERRE-PAUL PROULX

A region state must be small enough for its citizens to share certain economic and consumer interests but of adequate size to justify the infrastructure communications and transportation links and quality professional services necessary to participate economically on a global scale.

—Ohmae, 1993, p. 80

To present the framework I am developing to explain economic development of city-regions in North America, I have drawn on analyses of economic integration, international trade, economic growth, regional economics, industrial organization, economics of technological change, and sociology.[1] My objective is to understand the determinants of the growth of cities as they exist today and postmodern cities. The framework helps to interpret the evolution of cities from the industrial metropolis of the 19th century, based on factory production, to the modern metropolitan center of post-World War II, characterized by the central city serving as an office center and by population, retailing, and manufacturing activities moving to the suburbs. Rather than address that issue in this chapter, my purpose is to explore the changes currently modifying the modern metropolitan center and to better anticipate the outlines of the postmodern city.

Coffey (1994), who uses the terms *modern metropolitan center* and *postmodern city,* has recently prepared a monograph on the evolution of Canada's metropolitan economies, in which he identifies the following major factors underlying the emergence of the postmodern city:

- The displacement to newly industrialized countries of mass-production industries (and some back-office functions) followed by growth of service activities in cities
- The transformation of urban cultural and leisure activities from pastimes into economic activities
- "Sense of place" and its role in corporate investments and location decisions
- The emerging role of information and communications technologies in supporting processes of organizational restructuring associated with the globalization of production and distribution
- The commodification of information and the increasing "informationalization" of the urban economy
- A new wave of suburbanization of high-order office functions

As Coffey indicates, these are all related to the ongoing transformation of industrial economies. He also identifies a number of schools of thought on this subject.

Some scholars note that this transformation is modifying the notion of the firm as an organization into one of an organization of firms. They write about "enterprise webs" and a new environment in which world cities and trading blocs are replacing nations as the basic geoeconomic unit (see opening quote). Coffey (1994) sees metropolitan economies regarded as the relay points of national economic space, thus assuring the connections between the national economy and the global economy, on the one hand, and between the national and global economies and the regions, on the other.[2] He also notes that a transformation of metropolitan economies may be under way, one in which the downtown corporate activities complex, increasingly decentralized toward the metropolitan periphery in North America, is being replaced by a new economic base: tourism and cultural activities.

The framework presented in this chapter emphasizes some of these factors and incorporates others. The major elements of the framework are economic integration caused and accompanied by technological and organizational change in transportation and communications; the growing importance of services and information, a phenomenon related to the growth of intrafirm and intraindustry trade; the effects of location (place) on firms; the cumulative nature of economic growth and decline; intraregional and interregional networking; and internal and external economies of scale (externalities). I suggest that forces at play are changing the economic space of city-regions. These forces are progressively modifying the origins of imports and the destinations of imports and are causing a growing percentage of trade in services and products incorporating high technology to occur between cities and regions and to be decided in the urban context. I also anticipate growing disparities between cities and regions, the result of the strong effect of externalities and agglomeration economies on city-region growth. The following framework is subject to much further analysis and development. At this stage, the framework is eclectic, given the multiple sources of knowledge on which it draws.

Integration: Its Causes and Effects on City-Regions

Every day brings new articles, books, and working documents on integration. (For my own contributions to this literature, see Proulx, 1991, 1993.) Driven by technological change, business reorganizations, intergovernmental trade liberalization agreements, domestic policy changes such as deregulation and privatization, the growing services orientation of economies, converging consumer preferences, growing similarities in resource endowments in a service-oriented economy, and so forth, the mobility of goods, services, capital, people, and information in its multiple forms is increasing interregionally and globally. Integration is diminishing the importance of political borders and modifying the economic space relevant to firms, institutions, and governments, the majority of which are located in cities. A new technological paradigm based on generic technologies in electronics, information science, robotics, fiber optics, and ceramics, in addition to shorter product and service life cycles with the accompanying increase in the risk and uncertainty of research, is driving firms to network, to form alliances, and to undertake other organizational modifications that have significant implications for the role of city-regions in integrated economies. As locations for local-regional networking, information and technological exchanges, and stepping-off points for trade, city-regions are becoming international cities. Space and proximity, aspects neglected by most economists, are important in explaining the growth and decline of city-regions because externalities and agglomeration economies arise in such instances, and "place" has effects on firms.

Both increasing globalization and growing regionalization are occurring because significant international trade is taking place in blocs of countries. Regional and multilateral trade liberalization agreements are only one of many causes of this. Integration is causing the recomposition of economic space and the growth in importance of cities.

Industrial Location, Space, Firm Behavior, and Networks of Cities[3]

The development of cities in advanced economies does not correspond to central place theory developed in the 1930s and 1940s by Christaller and Losch. The reduction in transportation costs and consumer demand for variety have diminished the effects of geography and gravity of those models. Technological changes in transportation and communications, localization economies, and vertical and horizontal linkages between both local and distant firms have given rise to specialized cities and interregional trade. Recent work on organizational decisions of firms, particularly on firm networks, contributes to a better understanding of networks and hierarchies of cities. Examination of new technological and commercial organizational forms of cooperation—joint ventures, consortiums, and partnerships—helps explain the development of city-regions. Useful also in understanding networks of cities are the analysis of new behavior patterns of firms with models that are intermediate between hierarchies found in firms and markets and an examination of transactions costs and their role in the shaping of organizations.

A fruitful method of analysis of firm behavior involves examination of the extent and nature of firm networks for technological, financial, and marketing purposes. Networking between local and increasingly distant firms is undertaken to achieve economies of scale and to access and control technologies and complementary assets. This approach involves taking city-regions into consideration because head offices of firms involved in networking are usually located there. Some authors contend that a new hierarchy of cities is being established on the basis of the decision-making power (hence the number of head offices) found therein (see Bonneville, 1994). Intrafirm and intraindustry trade, with the accompanying foreign direct investment, networking, and alliances and trade in producer services, is continuing to grow in importance. Head offices and related services to firms operating interregionally and internationally are concentrating in city-regions to access the human resources, the innovation, and the multimodal infrastructures necessary for competitive and successful operations between city-regions of different countries and continents.

As indicated by Landesmann and Petit (1992), factor incomes (primarily returns on investment abroad but also returns on labor used abroad and governmental transactions) and international payments for certain specialized services (finance and insurance transactions, various business services, and payments on intellectual property) have risen significantly, a phenomenon explained by the reorganization at a regional and world level of manufacturing production and trade in goods. This result underlines the interdependence between goods and services. Network externalities that grow with additional users of accounting, telecommunications, financial, and insurances services lead to intraservices industry trade and specialization, much of which originates and flows through city-regions in particular because of the information technology infrastructure and human resources found there. The diffusion of information technologies has increased the potential for economies of scope in a number of service areas, leading to more diversified services and to high concentration in central business districts of certain cities, on one hand, and internationalization and regionalization of trade flows, on the other. Landesmann and Petit note that if economies of scope originate in technological production advantages, they will give rise to trade, whereas if they originate in organizational advantages of designing and producing a spectrum of differentiated products and/or services, they will lead to internationalization of production by affiliates rather than trade.

Local provision and concentration of high-order services in central business districts involve reductions in transaction costs, permit greater use of standardized procedures and managerial practices, and reflect particular systems of labor relations, language characteristics, national legal framework, and social relations. Initially, in such situations, trading would be more efficient than setting up affiliates or networking in different ways. But as a result of the internationalization of production and cross-national standardization of practices and demand structures, transaction costs will fall, as will the setup costs of international production. To help cities benefit from regionalization and globalization, one would analyze and track initial high trade flows, direct international investment, and networking factors.

Needed are analyses that incorporate internal and external economies of scale, transportation and communications costs, externalities and agglomeration costs and benefits, the local-regional milieux of information and technological cooperation and exchange, and the differences in location decisions between goods and services. (The production and consumption of goods, like that of high-order services, is more easily separable in space; that of low-order services is simultaneous, hence the location of suppliers near customers.)

Table 8.1, adapted from Camagni (1992), indicates the three organizational models of location that explain business location decisions and their relationship to urban-regional systems. The territorial model is geography bound, focused on production, and closer to the Losch and Christaller models. It applies more directly to the location of agricultural activities, public administration activities, and public and private services to persons. It overestimates the role of transportation costs, neglects input-output vertical relations between city-regions of similar rank, and does not consider network externalities, which come from city-region partnerships in networks of cities. The competitive model of business location reduces the role of transportation costs and hence breaks away from dependence on the local market. Competitiveness, marketing, and intrafirm and intraindustry trade are highlighted as features of firms and their activities. Space and heterogeneity between cities and regions are exploited, and the location of activities is determined by a series of geographic and historical factors (more on this later). According to the network model, innovation and networking become crucial features of firm activities, and interregional links diminish to some extent the importance of geographic space as an explanatory variable of structure, behavior, and performance of firms.

Location and space do remain important in this network model because in locations rich in technological and industrial culture (Alfred Marshall's industrial districts), firms become efficient partners with specific advantages for networking. In such locations (city-regions) also, firms have access to nubs of global and regional networks of technological, financial, and commercial information. The results are specialized mission units, initially localized near head offices and their historical location, usually near large metropolitan areas. Later, these units can relocate to more distant locations. The importance of understanding business location for a proper explanation of city-region growth and decline is obvious if one accepts that this approach has some explanatory power. According to this paradigm, one can expect to observe several types of networks of city. The first type involves complementary, specialized cities and regions that obtain market access and economies of scale and agglomeration through participation in the network. A second type is based on cooperation. It involves financial centers linked, for example, through telecommunications infrastructures, similar urban functions (such as tourist cities), and transportation connections. A third category is one in which firms from different locations network to supplement their respective strengths. For example, an innovation network might be formed to create a regional strength in research or to present a compelling argument for funding from higher levels of government for infrastructure investments.

Systems of cities are defined as systems of horizontal nonhierarchical relations between complementary or similar places, giving rise to

TABLE 8.1 Three Models of Spatial Organization

ORGANIZATIONAL MODEL	*TERRITORIAL*	*COMPETITIVE*	*NETWORK*
At the firm level:			
Nature	Firm aims at local market	Firm involved in international market	Network firm
Crucial function	Production	Marketing	Innovation
Strategy	Control local market	Control market shares	Controlling innovation
Internal structure	Unique location	Specialized functional units	Integrated functional units
Barriers to entry	Distance	Competitiveness	Continuous innovation
At the urban system level:			
Basic principles	Domination	Competitiveness	Cooperation
Structure	Vertically integrated (Christaller)	Specialization	Network of cities
Sectors	Agriculture, public administration, traditional service activities	Industry, industrial districts speciali-zation by network	High order services
Efficiency	Economies of scale	Vertical/horizontal integration	Network externalities
Policy strategies	None: Size of city determines its function	Initially none: Export base determines growth; currently: Reinforce competitiveness of each center	Intercity cooper-ation, building of network infrastructure
Objectives of interurban cooperation	None, other than military or diplomatic	Interurban division of labor	Economic, technological, and infrastructure
Network of cities	Hierarchical network	Complementary network	Synergistic network and innovation network
At the city level:			
Nature	Traditional city	Fordist city	Information city
Form	Internally relatively homogeneous	Specialized zoning	Multifunctional zoning and polycentric cities
Urban policy objectives	Power and image	Internal efficiency	External efficiency and attraction
Symbols	Palace, cathedral, market	Smokestack chimney, skyscraper	Airport, market

SOURCE: Translated and adapted from Camagni (1992), pp. 30-31.

externalities or specializations, complementarities, synergies, cooperation, and innovation. Location of economic activities in space, would, according to this approach, be organized according to different models, hence the need for an eclectic approach—sometimes corresponding to the hierarchical Christaller model, sometimes influenced by heredity and history, a theme stressed in new growth theory, as I indicate below.

Camagni, Boeckhout, and Pompili (1991), to explain the evolution of interregional disparities in Europe, combine different approaches into an eclectic view. They believe that the evolution of interregional disparities (some of which reflect the growth and decline of cities and towns) in the European Community (EC) depends on four main processes that occur on different time scales:

1. Long-term processes that are neoclassical and based on stages of development: This approach posits that a secular trend exists to reduce disparities through information diffusion, imitation processes, interregional movements of labor and capital, interregional spread of infrastructures, and policy intervention aimed at interregional disparities. These long-term processes imply diminishing disparities between city-regions.
2. Medium-term processes tied to waves of economic development and technological transformation emphasized by the theory of technological spatial diffusion of innovation and by cumulative neoclassical approaches: These processes depend on clustering of innovations and investment decisions at specific times and on the cumulative nature of many economic processes. According to this approach, radical innovations and advanced infrastructures are more rapidly adopted and developed in advanced and coreeas possessing better labor skills and higher demand density. This foreshadows growing disparities between city-regions as integration proceeds and underlines the importance of place-cities as determinants of firm growth.
3. Short-term processes of a cyclical nature, both exogenous and endogenous: Core regions are more capable of enduring these because of their stronger economic fabric.
4. Institutional decisions: Recent examples include the formation of the EC and, in North America, the Free Trade Agreement (FTA) and the North American Free Trade Agreement (NAFTA).

Many of these propositions are empirically testable. I have not yet undertaken this research task.

Interrelationships Between Industrial, Technological, and Spatial Dynamics: New Theories of Economic Growth and Their Application to Cities[4]

Previous sections have presented different theories to explain the location of firms and the effects of location on firms. Recent analysis expands on the effects of place on the development of firms and adds another essential dimension to the eclectic model of city-region growth examined in this chapter. Krugman's (1991b) model shows how increasing returns give history a determining role in the explanation of the geographic location of economic activities. Krugman also indicates that increasing returns are a pervasive influence on the economy and that increasing returns play a decisive role in determining the economic geography of different continents. His work is related to that of Romer (1986) and Lucas (1988). The essence of these new endogenous growth theories can be indicated as follows:

- Technological progress and growth are endogenous.
- Increasing returns and cumulative growth are characteristic.
- Human capital has a significant role to play in explaining economic growth.
- Unlike neoclassical growth theory, there is no convergence in growth between countries.
- Multiple equilibriums can exist.

Lucas demonstrates how an economy initially characterized by poor endowments of physical and human capital can remain dominated by a better endowed one. This is the origin of my hypothesis that the FTA and NAFTA will result in an increase in city-region disparities in North America, with the better endowed city-regions gaining ground on the poorer ones. Krugman and Lucas also make use of Myrdal's cumulative and circular causality concept in explaining how economic forces interact and accentuate initial trends in an economy.

These models enhance the understanding of the development of cities and regions because they have been applied (Krugman, 1991a) to explain regional development in national economies. Krugman uses this approach to explain how the U.S. manufacturing belt persisted so long by calling on the interaction of increasing returns and Marshallian pecuniary externalities. These models help to explain the concentration of particular industries in particular areas, such as microprocessors in Silicon Valley (California), traditional watches in Switzerland, and microtechnics in Baden-Württemberg (Germany).

Krugman (1991a) writes that "if there is one single area of economics in which path dependence is unmistakable, it is in economic geography; the location of production in space. The long shadow cast by history over location is apparent at all scales from the smallest to the largest" (p. 9). According to Krugman, the persistence of the U.S. manufacturing belt and a center-periphery model are the consequences of the interaction of increasing returns, transportation costs, and demand. Once established, a city-region reinforces itself through the interaction of the factors in a circular causation and cumulative Myrdal model.

In Schumpeter's analysis, historical accidents and significant shocks also have important effects on the growth paths of cities and regions. Krugman (1991a) presents a stylized story of the shift to the West (the slow start but subsequent growth of California and Los Angeles), in which expectations also play a role. He demonstrates that the final equilibrium results from the willingness of firms to invest, which depends on their expectations concerning the investment behavior of other firms. As Krugman writes, "the rise and decline of individual cities, and perhaps of somewhat larger regions, may indeed sometimes be the result of self-fulfilling optimism and pessimism" (p. 122).

Krugman implicitly recognizes a factor discussed and developed by European colleagues of *Groupe de recherches européen sur les milieux innovateurs* (GREMI), namely, that the proximity of firms facilitates synergies from which spring technological innovation. Firms permeate the environment in which they locate, and the growth of cities and regions cannot be understood without recalling history. Territory and space influence the development of firms, a recognition of the feedback relations between firms and the place in which they locate. Space becomes endogenized in the explanation of city growth. The dynamics of city-region growth cannot rely solely on explaining location decisions by firms but must incorporate an understanding of the functioning of local-regional milieux, which plays a significant role in the theory of city growth that is developing (see Proulx, 1991). History and location play a definite role in these attempts to explain city-region growth and decline. The role of increasing returns and circular self-reinforcing mechanisms (up or down) emphasize the importance of history in addition to technology, business strategy, intergovernmental agreements, and integration in explaining the development of North American city-regions.

Externalities and Growth in Cities[5]

Technological spillovers within cities and regions and between them in networks play a significant role in the explanation of city growth. Recent theories of growth, including those of Romer, Porter, and Jacobs, have stressed the role of technological spillovers, most of which occur in cities as a result of a close environment and interactions between people. This is the *milieux synergétiques* that GREMI colleagues have been emphasizing and that I have incorporated into my thoughts on city-region growth. Certain authors, especially Marshall, Arrow, and Romer, emphasize the importance of knowledge spillovers between firms in an industry. They conclude that concentration of an industry in a city should promote the growth of that city. These authors also predict, as does Schumpeter, that local monopoly is better for growth than local competition. Porter agrees with the intraindustry spillover hypothesis and the implication that regrouping and specialization in a city-region should help growth, but he believes that externalities are maximized in cities with geographically specialized competitive industries. Jacobs contends that knowledge transfers occur between industries and that as a result, variety and diversity of geographically proximate industries rather than geographical specialization promote innovation and growth. These theories explain both why cities form and why they grow. These theories are different from the more standard location and urbanization theories that address the formation and specialization of cities but not city growth.

A test of these theories by Glaeser, Kallal, Scheinkman, and Shleifer (1991) indicates that city-industries (such as autos and insurance) grow slower in cities in which they are more heavily overrepresented, a conclusion that does not support the Marshall, Romer, Arrow, and Porter results. They also found that industries grow faster in cities in which firms in those industries are smaller, a result that supports the views of Porter and Jacobs that local competition promotes growth. Glaeser et al. also concluded that city-industries grow faster when the rest of the city is less specialized, a finding that supports Jacobs. Another result agreeing with Jacobs's work is that specialized cities grow slower (weak evidence, according to Glaeser et al.). A significant association between a city's growth and competition appears to be lacking. Thus, there are significant focus and interest in externalities of different forms in attempts to explain city-region growth. Proximity, external scale economies due to urbanization, and high local demand are among the explanatory factors in their writings.

Conclusion

Although complex and multifaceted, the aforementioned factors do not explain in a satisfactory way the growth of cities. The challenge remains to formulate this framework in a testable way and to derive the policy implications that flow from it. The simplified model I have applied to policy-oriented projects on regional and urban development emphasizes the following factors among those discussed:

- Economic development in the countries' regions and cities to which local firms export and from which they import
- Spillovers and linkages between city-regions via foreign direct investment, multinational firms, and networking
- Horizontal and less selective public policies with local-regional governments being much more involved in the areas of support for innovation, human resources training, the provision of multimodal infrastructure, and paradiplomacy (i.e., networking with other cities)

The ultimate objective is to explain the functioning of a hierarchy of networks of cities, as Camagni has discussed in his works. Within each network, relations are horizontal. Between these networks, one observes market linkages toward the top, as cities in lower networks buy specialized products from cities in high-order networks, to which they sell certain inputs. Between these networks is also a manifestation of hierarchical linkages toward lower networks, as in the Losch model.

My framework implies that cities will attempt to expand their economic space by serving as gateways to the world for themselves and their hinterland. This approach also suggests that cities should strive to develop high-order functions and to promote activities of an international nature. The framework also proposes that because of the significant effects on large city-regions, cities should involve themselves in strategic planning and networking to benefit from the effects of integration.

NOTES

1. This chapter is concerned primarily with explaining the growth of large urban agglomerations. Linkages to their suburban environment do play a role in the explanation, but no effort is made to address the growth of local rural economies. The interested reader may consult Kusman (1994) for a review of selected empirical literature of interest.

2. See Proulx (1991) for an article in which these concepts play important roles and for references to the work of such colleagues as Soldatos (1991) on gateway cities.

3. This section draws heavily on Camagni (1992).
4. This section is based to a large degree on Krugman (1991a, 1991b) and Lecoq (1993).
5. The preparation of this section is based on Glaeser et al. (1991) and Camagni et al. (1991).

REFERENCES

Bonneville, M. (1994). Une revue des recherches sur les villes et l'internationalisation [A review of research on cities and internationalization]. *Revue d'Economie Régionale et Urbaine, 2,* 133-158.

Camagni, R. (1992). Organisation économique et réseaux de villes [Economic organization and networks of cities]. In P. H. Derycke (Ed.), *Espace et dynamiques territoriales* (pp. 25-52). Paris: Economica.

Camagni, R., Boeckhout, S., & Pompili, T. (1991). *Interregional disparities in the European community: Structure and performance of objective 1 regions in the 1980's.* Brussels, Belgium: Commission of the European Union, Directorate-General XVI.

Coffey, W. J. (1994). *The evolution of Canada's metropolitan economies.* Montreal, PQ: Institute for Research in Public Policy.

Glaeser, D. L., Kallal, H., Scheinkman, J. A., & Shleifer, A. (1991, March). *Growth in cities* [Mimeo, 3rd draft].

Krugman, P. (1991a). *Geography and trade.* Cambridge: MIT Press.

Krugman, P. (1991b). Increasing returns and economic geography. *Journal of Political Economy, 99,* 483-499.

Kusman, D. (1994, March). *Factors associated with the growth of local and regional economies.* Washington, DC: U.S. Department of Agriculture, Economic Research Service.

Landesmann, M. A., & Petit, P. (1992). *Trade in producer services, international specialisation and European integration* (Working Paper No. 9217). Cambridge, UK: University of Cambridge, Department of Applied Economics.

Lecoq, B. (1993, March). *Contribution àl'analyse des dynamiques industrielles localisées* [Contribution to the analysis of local industrial dynamics] (Document de travail No. 2/93, Université des Sciences et Technologies de Lille). Lille, France: CESURE.

Lucas, R. E. (1988). On the mechanics of economic development. *Journal of Monetary Economics, 22*(1), 3-42.

Ohmae, K. (1993). The rise of the region state. *Foreign Affairs, 72*(2), 78-88.

Proulx, P.-P. (1991). Cadre conceptuel et éléments théoriques pour l'analyse de la localization des activités économiques: Le cas des villes internationales [Conceptual framework and theoretical elements for the analysis of the location of economic activities: The case of international cities]. *Canadian Journal of Regional Science, 14*(2), 257-274.

Proulx, P.-P. (1993). Quebec in North America: From a borderlands to a borderless economy: An examination of its trade flows with the U.S.A. at the national and regional levels [Special issue on Quebec in the continental economy]. *Quebec Studies, 16,* 23-39. (Also available as Cahier 9327 [1993, August] of the Université de Montréal, Département des Sciences Economiques)

Romer, P. M. (1986). Increasing returns and long run growth. *Journal of Political Economy, 94,* 1002-1037.

Soldatos, P. (1991). *Buffalo as a gateway city* (Working paper). Montreal, PQ: Institute for the Study of International Cities.

Part IV

Case Studies of North American Municipalities

9 Planning a U.S.-Mexican Bi-National Metropolis: El Paso, Texas-Ciudad Juárez, Chihuahua

SAMUEL SCHMIDT

Introduction

The U.S.-Mexican border region has been growing in an uncontrolled fashion for decades, losing development opportunities and competitiveness in the process. Rapid migration and neglect from both federal governments have created poor border cities that struggle to attract investment. On the Mexican side, the infrastructure is built and rebuilt systematically, requiring tremendous amounts of money. Tijuana is starting its Urban Activation Plan with a 500-million-peso price tag (about $151 million U.S.), which includes building new highways and paving existing streets. Ciudad Juárez is starting a new governmental low-income housing development, building about 418 houses with an approximate cost of 6,500 pesos per unit plus 6,000 pesos per lot, including services with a total cost of $1,658,729. The 1,200 hectares expropriated by the city in 1993 created a territorial reserve sufficient to solve the present housing deficit of 47,000 houses. On the U.S. side, irregular settlements are sprouting. In El Paso, 15% of the population lack water and public services.

Since the inception of the Mexican National Border Program (PRONAF) in 1961, followed by the Border Industrialization Program (PIF) in 1965 (Küsel, 1988), which established the legal framework for creating

AUTHOR'S NOTE: Many thanks for valuable comments by Norris C. Clement and for assistance in collecting information and interviewing public officials in El Paso provided by Melinda de la Isla and Monica Lopez and by Maribel Limongi in Ciudad Juárez. The research was conducted with a grant from the Ford Foundation.

the *maquiladora* program, the regional economy was integrated into the international markets with little local economic diversification or improvement of the quality of life—in fact, some people claim the opposite has occurred because few resources were created for local development and for coping with tremendous challenges motivated by federal decisions.

Globalization is challenging countries and cities to be more competitive and to respond to global opportunities. Border cities are no exception. Their challenge is twofold: to advance from backwardness and to rapidly adopt new technologies and products to begin to capitalize on opportunities resulting from the North American Free Trade Agreement (NAFTA) and the global economy. For example, Dillinger et al. (1992) suggests that one of the challenges for Ciudad Juárez is to move on to higher productivity manufacturing and a more diversified economy.

The city is considered an economic, political, or cultural space. Experts agree that the city must provide economic viability, which includes minimal standards of quality of life. Quality of life in Canadian, U.S., and many European cities implies the availability of basic services (see Calavita, 1993),[1] but the term, as applied to a bi-national metropolis such as El Paso-Ciudad Juárez—which lacks those basic services and infrastructure and is plagued by the unsolved problem of street vendors and beggars—requires reformulating the concepts of urban planning, progress, and development.

Border urban planning has been conceptualized as regulation of urban zoning or application of good government, perhaps because city planners must keep in mind the need to satisfy short-term basic needs in the framework of rapid urban growth. Frequently, planning ends at the borderline, assuming a blank map in the other side of the line, although problems go beyond the physical border, as in the cases of the watershed and the airshed. Plans are often merely projections that do not address a complex border reality. Needless to say, no global or strategic planning has been carried out.

Conventional solutions to elevate quality of life propose increasing services. Although this approach can be right in the short term, it might not be sufficient to make the cities competitive because of the complex nature of local problems resulting from the complicated situation of two asymmetrical societies, in which pressures created in independent bi-national processes are combined in one urban setting. The incremental approach can create a new set of problems or make existing problems more acute by ignoring them. The search for nonconventional ideas such as bi-national planning is needed to provide a new developmental framework and mind-set.

Many people might oppose bi-national planning on the basis of the existing asymmetry between the United States and Mexico, but the idea is worth considering if for no other reason than because it provides a new way of thinking about U.S.-Mexican relations and border integration.

Most border cities are integrated environmentally and economically on a limited basis. Different initiatives facilitate exchanges, and a myriad of institutions are trying to integrate border cities. The Pan-American Health Organization (PAHO) is conducting a sister cities program to attack bi-national health problems. The University of Texas at El Paso (UTEP) and San Diego State University (SDSU) are trying to integrate educational programs with Mexican universities and school districts. Communication initiatives are developing in various spheres, including chambers of commerce and city officials such as those in Tijuana-San Diego, who are planning a new communication system (Rotella, 1993). But those initiatives lack formal integration and the will to incorporate bi-national competitiveness and often face legal and political obstacles.

In this chapter, I will characterize the U.S.-Mexican border region and some of the causes of border growth. I will also define the Paso del Norte region, explore planning deficiencies, review the obstacles for planning a bi-national metropolis, and propose the creation of a bi-national urban planning institute as an instrument for regional development and competitiveness.

Characterizing the Border Region

The U.S.-Mexican border region was created as a result of two modes of conquest—the Spaniards and the Protestant pilgrims—and a later expansion process in which half of Mexico's territory passed to the United States (Griswold del Castillo, 1990). Border disputes continued into the 20th century over El Chamizal, a strip of land between El Paso and Ciudad Juárez, as a result of a shift in the course of the Rio Grande.[2]

The border region historically has been perceived as empty space.[3] For the Spanish empire, the border did not have enough resources and labor for tribute. Scattered missions and *presidios*[4] helped create a Spanish presence motivated by a religious interest but never represented an economic interest. For the English colonizers, the frontier's wild west was an area for adventure and expansion: "Until the late 1800s the borderlands were regarded as badlands or *malpaís*—the realm of rattlesnakes, *bandidos,* and desert rats" (Barry & Sims, 1993, p. 19).[5]

The arid desert climate helped perpetuate the perception of the area as "wastelands without beauty or life" (Barry & Sims, 1993, p. 19) and empty space. This perception facilitated the establishment of the highly polluting *maquiladora* industry, with some 2,000 plants and 500,000 employees, and is being used as a rationale to install toxic and nuclear waste dumps in the border region.

In both countries, the border region adopted an industrialization process developed in areas with an abundance of natural resources, mainly water. This industrial model was applied without ecological adaptation, depleting and damaging scarce resources, a process that by now has resulted in a catastrophic environmental scenario with little appeal for new investors, who may find the border a hindrance instead of an advantage. Paradoxically, border industrialization and its effects are now one of its most important obstacles in attracting new investment, reducing the area's competitiveness.

The border cities' economy has not been expanding in an orderly fashion because the cities were not prepared for numerous migrants or for the establishment of a large number of industrial plants. Consequently, these cities were unable to develop the infrastructure to integrate the new industrial parks into their own economies. The *maquiladoras* imported advanced technology that was never transferred to local businesses. Few of the *maquiladoras* incorporate more than 2% of local components; thus, they became industrial islands that put significant pressure on a limited infrastructure, producing a tremendous environmental impact.[6]

Paradoxically, thanks to the *maquiladoras,* the U.S.-Mexican border region can be considered the most globalized region on the continent but is unable to use this advantage for further development because of imbalances created by *maquiladoras.* Lack of integration into the local economies prevents regional competitiveness. For the U.S. economy, *maquiladoras* represent a source of cheap labor to reduce production costs but do not create other economic linkages or develop economic potentials at the border. For Mexico, excluding salaries (which are basically minimum wage), they bring little economic benefit and limited additional economic opportunities.

National neglect of the border region is reflected in its little or no role in national decision making,[7] although the border's function is to regulate trade and migration. In the United States, border problems apparently are not considered of a national dimension, and, consequently, the federal government does little to solve them. The *maquiladora* program begun in 1965 is perhaps the best example. It was aimed at stopping migration, which was regarded as a national problem. Thus, no one was prepared to cope with the resulting infrastructure and environmental problems because these

TABLE 9.1 Income at U.S. Border Cities (in U.S. Dollars, 1990)

	Median Family Income (Dollars)	*Per Capita Income (Dollars)*	*Below Poverty Level (Percentage)*
National (United States)	35,225	14,420	13
Texas	31,553	12,904	18
Brownsville	16,889	6,284	44
McAllen	23,809	9,814	33
Laredo	19,910	6,981	37
Eagle Pass	14,397	5,720	46
El Paso	25,157	9,603	25
Arizona	32,178	13,461	16
Nogales	20,386	7,795	31
California	40,559	16,409	13
Calexico	19,872	6,595	32
San Diego	39,318	16,401	13

SOURCE: U.S. Census Bureau.

were supposedly Mexican problems. Although the goal of stopping migration was never achieved, border cities developed large areas with substandard housing and public health problems because pollution does not respect political boundaries.

Mexico's goal was to create border development to retain its population. National and international linkages leave little local benefit at the border region, creating islands of poverty and resentment among the border population. Mexico needed to create free trade zones to satisfy border needs (Secretaría de Hacienda y Crédito Público, 1980) and defuse political protest in the face of scant opportunities for development. Regarding development, however, the expected economic impact was irrelevant.

The in-bond industry attracted U.S., Asian, and other countries' plants, but it did not create a bi-directional flow. The *maquiladoras* received components for assembly in Mexico and exported assembled products back to the United States. This type of economic growth has put border cities in a disadvantaged position. With the exception of San Diego—which has a historic involvement with the U.S. Navy, called by Clement and Zepeda (1993) the "economic growth machine" (p. vii) supporting the most important navy base on the West Coast—border cities are among the poorest in the United States. The percentage of people living in poverty is higher than the national average, and the average per capita income is lower than the national average (see Table 9.1).

TABLE 9.2 Labor Force Status in U.S. Border Cities: Annual Averages, 1990

	Civilian Labor Force (Thousands)	*Unemployment Rate (Percentage)*
National (United States)	124,000.8	5.5
Texas	8,443.0	6.2
Cameron County (Brownsville)	104.8	11.7
Hidalgo County (McAllen)	164.4	19.1
Webb County (Laredo)	53.2	10.8
Eagle Pass	13.9	26.7
El Paso County	250.8	10.7
Arizona	1,726.0	5.3
Nogales	n/a	n/a
California	14,670.0	5.6
Calexico	n/a	n/a
San Diego County	1,174.4	4.5

SOURCES: Data for United States, Arizona, and California from U.S. Department of Labor (1991), *Employment and Earnings,* Vol. 38, No. 5, Table D-3. Data for Texas from Texas Employment Commission, Economic Research and Analysis Department (1990).

Unemployment on the U.S. side of the border is also higher than the national average (see Table 9.2). Many border cities double the national unemployment average and others multiply it a few times. For San Diego, statistics refer to the county, but the numbers for San Ysidro or Chula Vista would be similar to the rest of the border cities.

For Mexican cities, unemployment is lower than the national average, but the rate of economically inactive population is still high (see Table 9.3). Only in Nogales does the percentage of employed persons top 50%. One of the explanations for the lower border unemployment rate might be the Mexican method to measure employment, which results in low unemployment rates, added to people employed, documented or not, on the U.S. side. *Maquiladoras,* perhaps the most important employment source, also contribute to the higher percentage of employment. Nevertheless, border cities are poor and lack basic services. Although some people claim that *maquiladora* jobs pay at a rate higher than the national minimum wage,[8] it is also true that life on the border is more expensive.

The use of poverty by businesses creates a vicious cycle. A poor city attracts businesses interested in using cheap labor and taking their profits away. These businesses do not develop additional industry and pay little in taxes because the city grants tax breaks to attract them. The city is unable to develop a better infrastructure and improve quality of life and

TABLE 9.3 Mexico: Economically Active Population, 1990

	Population Over 12 Years of Age (Thousands)	*Employed (Percentage)*	*Unemployed (Percentage)*	*Economically Inactive (Percentage)*
National (Mexico)	55,913.8	42	3	55
Tamaulipas	1,610.2	43	2	55
Matamoros	217.0	48	2	50
Reynosa	202.8	45	1	54
Nuevo Laredo	154.7	45	1	54
Coahuila	1,397.4	42	1	57
Piedras Negras	69.9	46	1	53
Chihuahua	1,724.4	45	1	54
Ciudad Juárez	563.1	50	1	49
Sonora	1,293.0	43	1	56
Nogales	75.8	51	1	48
Baja California	1,170.6	48	1	51
Mexicali	427.3	47	1	52
Tijuana	525.9	50	1	49

SOURCE: Instituto Nacional de Estadística, Geografía e Informática (INEGI), 1991.

cannot become more competitive to attract better-paying jobs; consequently, it attracts corporations interested in hiring cheap labor.[9]

Infrastructure in border cities is insufficient. Local resources to meet present and future challenges are nonexistent; therefore, border cities rely on state or federal funding. Seventy percent of bilateral trade is by land ("Realizarán México, EU y Canadá," 1994). In 1992, almost 300,000 trucks crossed through El Paso and over 300,000 via Laredo, putting tremendous pressure on the limited existing infrastructure. The need for economic integration and trade via NAFTA will find the bottleneck in the border infrastructure difficult to overcome. The existing and future environmental impact will cause more degradation of the quality of life on the border[10] unless strong measures are taken right away. If terrestrial trade grows, existing infrastructure will be unable to support it, resulting in a loss of competitiveness. Growth in the border region can be at once an incentive for a larger economic scale and not an obstacle for manageable growth.

Only recently, because of NAFTA negotiations, both federal governments agreed to intervene in attacking the border environmental crisis, but their commitment is insufficient (Inter-Hemispheric Education Resource Center, 1993). The U.S. contribution was limited by Congress despite the *maquiladora* profits and taxes paid by the corporations that go to the

TABLE 9.4 Urban Population in Mexican Border States (Percentages)

	1940	*1950*	*1960*	*1970*	*1980*	*1990*
National (Mexico)	35.1	42.2	50.7	58.7	66.3	71.3
Border average	42.5	52.1	63.8	71.8	77.9	84.7
Baja California	50.5	64.5	77.7	84.3	85.3	90.9
Chihuahua	36.7	44.1	57.2	65.4	70.3	77.4
Coahuila	50.6	57.4	66.7	72.7	77.4	86.1
Nuevo León	43.9	55.9	70.4	76.5	87.4	92.0
Sonora	32.7	45.3	57.6	66.5	70.5	79.1
Tamaulipas	45.5	53.0	59.8	68.9	75.1	81.1

SOURCE: Lorey (1993), p. 22.

United States.[11] Mexico's contribution is higher because of the slight intervention of Congress in governmental expenditures and the higher availability of resources for the executive branch of government. The Mexican government has more discretionary funds to be used politically, as was the case with the funds allocated to support passage of NAFTA.

Governmental intervention can create a paradoxical situation. It can control urban growth, making the city more competitive, which can attract more population. The next section explores aspects of growth in the border region.

Population Growth in the U.S.-Mexican Border Region

In the latter half of the 20th century, population growth rates in border cities have exceeded their respective national averages (Herzog, 1986), creating a high urban concentration. In Mexico (see Table 9.4), Baja California and Nuevo León have more than 90% urban population, and in the United States (see Table 9.5), with the exception of New Mexico, urban population in the border states is above the national average. This growth is expected to continue. By the year 2010, Ciudad Juárez will contain 40% of the population of the state of Chihuahua, and Tijuana 50% of the population of the state of Baja California (Schmidt, Gil, & Castro, 1994).

Some of the growth on the Mexican side of the border originated with the *bracero* program, which facilitated the temporary migration of Mexican labor. It was also encouraged by the government, which used the slogan "to rule is to populate."

TABLE 9.5 Urban Population in U.S. Border States (Percentages)

	1940	*1950*	*1960*	*1970*	*1980*	*1990*
National (United States)	56.5	64.0	69.9	73.6	73.7	75.2
Border average	56.9	71.5	81.2	86.0	86.3	87.7
Arizona	34.8	55.5	74.5	79.6	83.8	87.5
California	71.0	80.7	86.4	90.9	91.3	92.6
New Mexico	33.2	50.2	65.9	69.8	72.1	73.0
Texas	45.4	62.7	75.0	79.7	79.6	80.3

SOURCE: Lorey (1993), p. 22.

During the years of the Mexican miracle (from about 1940 through 1970), the country recorded high rates of population growth. In the 1970s, the 3.5% average population growth concerned politicians, who put in place a program to reduce the birth rate (Godwin, 1977). Nevertheless, population continued to grow because of improved health care, increased life expectancy, and decreased infant mortality. This in itself put pressure on urban centers.

Urban growth was spurred by the agricultural policy. Vellinga (1993) says "the Mexican revolution sacrificed agriculture to industry and the countryside to the city" (p. 52). This trend continued with the import-substitution policy, which subordinated agriculture to industry. The peasants' response to decreasing economic opportunities was to move in considerable numbers from "high density areas and poor income to economically more attractive areas" (p. 53). Rapid urbanization was a consequence of industrialization. In 1940, Mexico's urban population was 35.1%; by 1990, it reached 71.3% (see Table 9.4). Border states, especially border cities, attracted high numbers of migrants. Urban growth was also intensive in the United States, increasing from 56.5% in 1940 to 75.2% in 1990. Border states and cities grew even faster (see Table 9.5).

Mexican rural migration to the cities started a process of concentration of urban poverty. The World Bank (1991) recognized that the different housing projects it supported helped exacerbate this concentration, creating acute urban problems. Existing networks in Mexican cities helped absorb migration (Lomnitz, 1987). Hiernaux (1986) shows that 56.8% of people who arrived in Tijuana were attracted by relatives or friends (pp. 90-94). A prevalent culture of poverty (Lewis, 1961) helped sustain political stability because the networks functioned as a support group. Eventually, Mexican cities were saturated, and migration to the United States served as a safety valve to alleviate urban unrest. Many Mexicans remained in border cities; others used those cities as transit stations. For the Mexican government, disorderly urban growth was a solution to intense migration.

The bracero program, begun in 1942, showed Mexicans the road to U.S. jobs (Herrera-Sobek, 1979). During its 22 years, 4.5 million Mexicans came into the United States for seasonal work (Vernez & Ronfeldt, 1991, p. 1189). Its termination did not motivate Mexicans to remain in Mexico, working at low-paying jobs. Although accurate statistics about the Mexican population in the United States are lacking, Diez Canedo (1984) reports a 1975 study estimating the presence of 8.8 million Mexicans in the United States. Also according to Diez Canedo, in 1976, some 781,474 Mexicans were considered deportable (pp. 30-31). Migration continued during the 1980s. The Immigration and Naturalization Service (INS) reports that as of February 13, 1992, around 5.3 million Mexicans legalized their migration status as result of the Immigration Reform and Control Act (U.S. Department of Justice, 1993, p. 17).[12]

Migration to the United States in the 1990s tended to increase. In 1992, the border patrol reported the detention of 1.1 million Mexicans. The patrol estimates that for each detainee, three cross the border. Adding these 3.3 million to 1.1 totals 4.4 million entering or attempting to enter. The INS modifies this number, reckoning that each person attempts to cross an average of 5 times before being caught. Dividing the total 4.4 million by 5 gives a total of 880,000 undocumented migrants, a number apparently accepted by Mexican migration authorities. Another analysis (Santibañez, Valenzuela, & Velazco, 1993) reports 2.8 crossing attempts for each detainee, which results in 1,571,428 undocumented migrants in 1992. This figure, plus 900,000 documented migrants the same year, brings the total number of Mexican migrants to more than 2.5 million, representing almost 3% of the total population of Mexico. In addition to this number, during 1990 to 1991, 1,625,235 migrated legally. Because not all of these migrants remain, it is difficult to know the real number of the Mexican population in the United States. Recent estimates place this figure around 1.5 million (Mendoza, 1994).[13] I concluded elsewhere (Schmidt, 1993b) that because this migration flow is a result of extreme poverty, which leaves Mexicans with no options in their own country either in the countryside or in the cities,[14] these migrants must be considered economic refugees.

Despite lack of agreement on the number of Mexicans who live in extreme poverty, estimates range in the tens of millions. The agency responsible for Mexico's food program (*Sistema Alimentario Mexicano*) in the 1980s estimated 20 million people living marginally (Luiselli, 1980, p. 22).

TABLE 9.6 Growth in Mexican Border Cities (Percentages)

City	*1940-1950*	*1950-1960*	*1960-1970*	*1970-1980*	*1980-1990*
National (Mexico)	3.1	3.5	3.8	3.9	2.2
Border average	4.4	4.7	4.2	3.6	2.4
Ciudad Juárez	9.6	7.9	4.5	2.9	3.8
Tijuana	13.8	9.8	8.0	2.8	5.0
Mexicali	13.2	10.5	4.2	2.6	2.5
Nuevo Laredo	7.2	4.9	4.9	3.1	.8
Nogales	5.9	4.4	3.3	2.3	4.9
Piedras Negras	5.8	5.0	-0.9	5.1	3.6
Matamoros	11.3	7.3	4.1	3.2	3.5
Reynosa	13.7	8.1	6.4	3.6	3.2

SOURCE: Data for national and border average calculated from Lorey (1993), pp. 25-38. Data for cities from García Amaral (1993), pp. 21-22.

Balboa Reyna (1991, pp. 4-5) estimates that 34 million Mexicans live under extreme poverty. The Economic Commission for Latin America (ECLA), using data provided by the Mexican government, reports 13.6 million people in extreme poverty in 1992, claiming an 8.7% reduction in this segment of the population from 1984 to 1992 (Sainz, 1993, p. 3). Local authorities in Ciudad Juárez estimate that 64% of the city's population lives below the poverty line.

Poverty forced Mexicans to move to the city—any city—and the border was no exception. Border cities grew faster than the national average[15] (see Table 9.6) with high migration rates in the 1950s and 1960s and a natural increase in the following decade (Arreola & Curtis, 1993, p. 29). In the 1980s, the national and border growth average decreased; nevertheless, the big border cities continued to grow at an above average rate.

U.S. border cities also grew faster than the national average (see Table 9.7) because of military bases,[16] Mexican migration, and the *maquiladora* program, which attracted corporations and additional Mexican migrants who remained in the border region after the 1960s.

Border military facilities also had an impact on Mexican cities, mainly because soldiers routinely crossed into Mexico for entertainment (Hiernaux, 1986). Ten of the Mexican cities studied by Arreola and Curtis (1993) are situated within driving distance of less than 1 hour to several hours from U.S. military installations. Another externality favoring rapid growth was the opening of border ports of entry, which creates an incentive for urban development and economic opportunities.

TABLE 9.7 Growth in U.S. Border Cities (Percentages)

City	*1940-1950*	*1950-1960*	*1960-1970*	*1970-1980*	*1980-1990*
United States	1.4	1.9	1.3	1.1	1.0
Border average	3.7	3.9	2.3	2.3	2.4
Brownsville	.6	3.3	.1	6.1	1.6
Calexico	1.9	2.4	3.3	3.6	2.9
Eagle Pass	1.3	6.6	2.7	3.9	9.6
El Paso	3.5	11.2	1.6	3.2	2.1
Laredo	3.1	1.8	1.4	3.2	3.4
McAllen	6.9	6.3	1.4	3.2	2.7
Nogales	2.0	1.8	2.3	7.5	2.4
San Diego	6.4	7.1	2.2	2.6	2.7

SOURCE: Calculations based on Lorey (1993), pp. 21, 49.

The disorderly agglomeration of people could cause enormous urban pressure, but for Mexico and Latin America, the culture of poverty has created housing solutions such as irregular settlements, which, although unsatisfactory, are helpful to avoid urban political instability. Irregular settlements were created with governmental complacency because they helped ease the demand for housing and at the same time provided a new link in the corporatist scheme of government. The Mexican government tolerated and sometimes supported a corrupt squatter's leadership in exchange for votes.

Similar innovative housing ideas started cropping up in U.S. border cities. Social solidarity, which produced overcrowding of dwellings, and cheap housing solutions helped to settle down migration with little pressure for the government but were also an incentive for disorganized urban growth. Although both governments provided low-income housing solutions, society created its own. In Mexico, these human settlements are called *colonias populares* or popular neighborhoods, a name apparently borrowed in the United States, where they are called *colonias.* Substantial differences exist between the two countries, however, in this respect.

In Mexico, people create irregular squatter settlements and start building houses with waste materials. Viewed in a positive light, this is in effect a recycling program because people use old cardboard, used tires, and discarded wood (Hiernaux, 1986). Through the years, when people have legal title, they start using more permanent materials such as brick or adobe.

This, of course, is not a new Latin American phenomenon, but the emergence of these constructs on the United States-Mexico border was an innovation. Now it is common to see *cartolandias*[17] (housing projects made of cardboard) spreading along border cities. In U.S. *colonias,* people buy land, bring in a recreational vehicle or mobile home, and, in a second stage, build a house around the vehicle.

Colonias are becoming characteristic of a bi-national metroplex such as El Paso-Ciudad Juárez because on both sides of the border, poor people cannot afford housing. In the United States, *colonias* are economically distressed areas (Rio Grande Council of Governments [RGCOG], 1993), with a *colonia* defined as an unincorporated community of five or more housing units, usually overcrowded and lacking water, wastewater facilities, and paved roads. Outdoor latrines are improperly installed, and septic systems often flood in heavy rainfall, increasing exposure to bacteria and other pathogens that cause disease. Financial resources are inadequate to satisfy minimal needs and services such as mail, transportation, and health facilities. Some *colonia* residents have private water wells, which are often adjacent to and contaminated by the flow from inadequate septic systems. Some people tolerate this condition because moving to a *colonia* allowed them to escape city violence; for others, it was the only chance to purchase property. *Colonias* create a need for immediate governmental action to solve social needs and to simultaneously use scarce resources for improving infrastructure to develop competitiveness.

■ The Paso del Norte Region

The Paso del Norte region is composed of the states of Chihuahua in Mexico and of New Mexico and Texas in the United States. The New Mexico portion includes Doña Ana County, containing the cities of Sunland Park, Las Cruces, Mesilla, and Hutch Village. The Texas portion includes El Paso County, containing El Paso, Socorro, Clint, Horizon City, Anthony, and Vinton Village. Total population on the U.S. side is 655,867. The Mexican side of the region includes the municipalities (counties) of Ciudad Juárez, Praxedis G. Guerrero, and Guadalupe, with only one city: Ciudad Juárez. Population on the Mexican side is about 940,000.[18]

The Paso del Norte region sits in the middle of the Chihuahua desert. Its climate forces people to live with air conditioning most of the year, raising the cost of utilities.[19] Water is scarce, and at present pumping rates, underground water will be depleted in no more than 50 years.

Infrastructure lags current economic needs, and the quality of education is substandard. The most important university is a teaching university struggling to take off via the creation of doctoral programs. These poor cities have little incentives for cultural development, and the arts are not sufficiently supported. Cultural and artistic disadvantages reduce the cities' appeal compared with other Texas cities.

The cities in the Paso del Norte region struggle against backwardness and reduced quality of life. According to the city of El Paso Human Services Department, as of May 1992, 26.8% of the city's population was living below poverty and 22.5% received food stamps. Some of the reasons for poverty in El Paso are high unemployment, lack of skills, low levels of education, and the border stereotype, which attracts corporations willing to pay low wages because the border region is considered inexpensive.[20] Unemployment in El Paso is about 10.7%.[21] Violence, illnesses, and environmental hazards threaten the city's inhabitants.[22] Instead of strategic planning, pressing immediate needs are a priority for local governments.

Attracting businesses is perceived as an instrument to overcome backwardness via job creation, although these are minimum-wage jobs. Seldom can a city become competitive with minimum-wage jobs. The need to satisfy basic needs makes the elevation of the standard of living and international competitiveness a secondary issue. In fact, the city of El Paso's sales pitch apparently relies on its inexpensiveness, the Ciudad Juárez's low wages, and an integrated bicultural setting.

Interaction between U.S. and Mexican border cities is strong to the extent that bi-national metropolises are considered an economic and environmental region. Some analysts estimate that Ciudad Juárez generates 20% of the jobs in El Paso and between 60% and 90% of retail trade in downtown El Paso (González, 1993, p. 1A). Other approximations reckon that 30% to 40% of retail trade in El Paso is generated in Ciudad Juárez.[23] Another study estimates that "sixty-four percent of trucks report making more than seven trips across the border per week . . . and shopping accounts for 27% of automobile trips and 36% of pedestrian trips into El Paso" (Goodman Corporation, 1992, p. ES-3). So far, however, with the exception of *maquiladoras* (which are not interested in the local economy), this has not been an argument strong enough to massively attract corporations.

Nevertheless, constant efforts are made by business and community leaders to create regional linkages following a strategy of regional development. Among relevant examples are the Camino Real project to inte-

TABLE 9.8 Border Crossings: El Paso-Ciudad Juárez

	1991	*1992*	*1993*
Cars	12,463,264	13,274,599	15,074,668
Persons in cars*	31,158,160	33,186,497	26,678,437
Pedestrians	5,217,274	5,186,795	6,355,050
Total persons	36,375,434	38,373,292	66,696,092
Average persons per month	3,031,286	3,197,774	5,558,067

SOURCE: U.S. Customs.
* 2.5 persons per car. U.S. Customs uses two figures, 2.5 or 2.8 persons per car. Goodman Corporation (1992) considers an overall automobile occupancy of 1.95 persons per vehicle.

grate Chihuahua, Southwest Texas, and New Mexico with tourism and the arts; Operation Good Neighbor, including business and government groups; and Industrial Development Council Briefings, sponsored by the El Paso Hispanic Chamber of Commerce to teach Mexican nationals how to do business in the United States and U.S. nationals how to do business in Mexico. In November 1993, the Greater El Paso Chamber of Commerce and the Ciudad Juárez Cámara de Comercio signed an agreement to promote tourism and the regional economy.

Numerous academic and cultural interactions occur in the region. In 1992, 10 universities in Texas, New Mexico, and Chihuahua created the Bi-national Consortium of Higher Education Institutions to exchange students and faculty and to create bi-national teaching programs. Family interactions are extremely common: Some families have members on both sides of the border. Consequently, border crossings number into the millions each year (see Table 9.8). On average, all the population in the region crosses twice a month, requiring a nonexistent integrated highway and transportation system.

Regardless of the multiple contacts, reality shows that functionally the cities are not integrated. Communication between city authorities is either superficial or nonexistent, and there is apparently no awareness of the need for joint planning.

Prospects for viable regional governance are slim at best under present conditions. A recent change in the two city administrations (El Paso and Ciudad Juárez) moved intercity coordination efforts many years back.[24] The numerous economic interactions are the result of competition and are not considered an advantage for regional development; consequently, there are no political initiatives to develop a regional approach to planning, not even one sufficient to take advantage of NAFTA-related opportunities.

Obstacles to Bi-National Planning

Any border is inherently an international area, but not all borders are competitive. Nationally or internationally, border cities must show their relative advantage and develop their competitive advantage.

Planning is an instrument to create the conditions to make the city competitive and to facilitate its development.[25] The main goal of planning is to enhance quality of life as a necessary condition for economic prosperity (Clement & Zepeda, 1993). Bi-national planning is an instrument to avoid conflict (Graizbord, 1986) and to better use regional resources, but national differences become hurdles for bi-national planning. Because starting a bi-national governmental effort will have to be sanctioned by both congresses, one of the most relevant planning obstacles for the United States and Mexico is diplomacy. This hurdle is second in importance only to the economic asymmetry between both countries.

The federal system itself, which both countries have, is another obstacle. Federal governments have agendas that might not coincide with state and local needs. The sovereignty issue may deter local border developments. One case is the proposed San Diego-Tijuana airport. The city of San Diego tried to move its airport to the borderline to take advantage of Brown Field and the Tijuana airport. The Mexican federal government rejected the project on the basis of a sovereignty rationale.

Another example is the U.S. migration policy and its corresponding border blockades enforced by the federal government, which have strong economic impact on the border region. Smith (1992) says that "the policy implications are obvious: We should stop focusing on immigration as primarily a law enforcement issue and recognize that long-term solutions can be found only in the context of the U.S. and Mexico's rapidly growing economic and social interdependence" (p. 126).

Bi-national obstacles to planning can be found at the state level because power arrangements favor one region over another. In the state of Texas, the Mexican American Legal Defense Fund (MALDEF) and the League of United Latin American Citizens (LULAC) sued the state government for discrimination against border education (Stoddard, 1992). The state of Texas is preparing itself for NAFTA by improving highway connections from Laredo to Canada, ignoring the state's southwest, which is closer to Canada and in a better geographic position to connect both the U.S. and Mexican west and east coasts. Perhaps this decision is based on Laredo's integration to Houston, Dallas, Austin, and San Antonio, the state's centers of power. Laredo is also linked to Monterrey, one of Mexico's most advanced cities, and to Mexico City via a high-quality highway.

City politics and the politics of city planning, which are influenced by local interests, also deter regional planning. Local conditions, such as fast growth and rapid change in land use, affect not only good city planning but also the perception of the neighbor's condition. Arreola and Curtis (1993) assert that "the popular perception of the Mexican border towns still conjures two persistent images. First, border towns are seen as tawdry yet convenient and accommodating tourist outlets, albeit for short term visits. Second, they are viewed as small places, not really large enough to be considered cities" (p. 3). In other words, Mexican border cities are not considered reliable partners. This prejudice, added to a lack of communication systems and good data,[26] prevents joint decision making.

Some planners, such as those in San Diego, used to work as if the neighboring city did not exist. Other planners, such as the ones in El Paso, acknowledge the need to link public services bi-nationally but do not always have a positive response from the neighboring city or from their own administration. The El Paso-Juárez Mobility Improvement Program (Goodman Corporation, 1992) prepared for the City of El Paso reports lack of personnel for international projects. The project also proposes to modernize transportation infrastructure for the entire region. Although part of the U.S. portion is well advanced, the Mexican side ran into profound complications between city and federal priorities and lack of financial resources.[27]

City planners from El Paso, Sunland Park, Las Cruces, Doña Ana County, El Paso County, and Ciudad Juárez created the Regional Planning and Development Committee (Goodman Corporation, 1992),[28] which practically disappeared when both El Paso and Ciudad Juárez administrations changed in 1993 and the priorities for the new mayors changed.

An additional obstacle is the lack of integration of the border cities into their national economies. With the exception of San Diego-Tijuana, although none of the border metropolises are fully integrated into their respective economy, national economic forces can stop the integration, hoping to incorporate the region into the national economy in the future. In all cases, the U.S. city is better incorporated into the national economy than the Mexican counterpart.

Local competition, admnistrative decisions, and cultural differences also impede regional and bi-national planning. Cities often compete against each other for the same scarce resources, even though some cities use the border to attract tourism. Ciudad Juárez and El Paso are in different time

TABLE 9.9 Bridge Revenues, Ciudad Juárez to El Paso (in Dollars)

Year	*Cars*	*Pedestrians*	*Total*
1991	11,572,100	2,283,794	13,855,894
1992	12,782,082	2,345,620	15,127,702
1993	12,618,876	2,918,281	15,537,157

SOURCE: U.S. Customs.

zones during the fall and winter. Only a limited number of businesses take advantage of the 1-hour difference, which for the most part is confusing and prevents the organization of joint activities. Different languages, symbols, and concepts of time and space[29] prevent good communication and understanding and obstruct bi-national coordination.

One of the most salient features regarding the two countries' governments is the difference in the tax base. In the United States, the city and county have tax powers, whereas in Mexico the city or *municipio* (county) controls only the property tax, which is insufficient for its needs because of a poor cadastral service and limited opportunities to increase taxes in an economically deprived city. In addition, large areas of Mexican border cities cannot be taxed because of the existence of irregular settlement. The other taxes are collected by the treasury department, sent to Mexico City, and partially returned to the states and cities.[30]

The Mexican federal government signs a yearly agreement with each state called *Convenio Unico de Desarrollo Social* (CUDS; Social Development Unique Agreement), which sets fiscal transfers from the federal to the state government. The state government in turn allocates a portion of the money to the city. The Salinas administration applied the 80-16-4 formula. Of every 100 pesos collected, the federal government retains 80, the state government 16, and the municipal government 4. The latter serves the people.

One example of the disparity between the two systems is the international bridges in El Paso-Ciudad Juárez (see Table 9.9). Two of the bridges are toll bridges; the third is free. El Paso claims a revenue of $7 million from its bridges. Mexico charges double the fee, but Ciudad Juárez keeps nothing because all the money goes to Mexico City. A potential investment of $14 million could make a big difference for Ciudad Juárez's infrastructure.

The Mexican tax system creates a strong political dependency. Because the transfer of money is decided in Mexico City, both state and city must lobby the federal government for economic resources, leaving most administration and development decisions in the hands of bureaucrats and politicians. Consequently, many decisions are based on discretionary money available in Mexico City with a political price. This may not have been a problem in the past for elected officials belonging to the *Partido Revolucionario Institucional* (PRI). Now, however, because the opposition is winning elections, as in the case of the two most important border metropolises (Ciudad Juárez and Tijuana), city authorities are limited by their political party positions and an arrangement of power that goes beyond the city and limits their lobbying capabilities.

In the United States, most local decisions are made on the basis of local arrangements of power, and development projects can be submitted to the federal government when money is available. Local taxes are limited because of the poor condition of border cities. Local taxes also have to fund school districts and counties. In Mexico, education is included in the CUDS.

This political condition makes bi-national planning difficult. A potential joint project involves the following two hypothetical situations: The Mexican city can lobby the federal government, but the U.S. city may not be that lucky; thus, it needs to use its own limited resources that might not be sufficient. In the reverse situation, the U.S. city can commit money for the future including the project in its plans, but the Mexican city may not have federal money in the future.

Another governmental barrier for bi-national planning is the administrative culture. In Mexico, every change of administration in every level of government replaces personnel across the board, including department heads and sometimes technical staff. Many times the staff leaving office might take the plans with them, creating an absence of institutional memory and waste of resources. "When the (Mexican) environmental agency was abolished, three thousand people left their desks and projects sitting there. Some actually threw plans in the wastebasket" (Brown, 1986, p. 42).[31] In the United States, changes are not expected to happen in the departments, and the plans remain,[32] although some of them are lost.

Another difference is that in the United States, public hearings are required for different issues. In Mexico, they may or may not be held. It is up to the political will of government to invite society to bring its opinion, and when it happens, it can be to legitimize a decision already made.[33]

Another potential obstacle to bi-national planning is the policy rationale. In Mexico, it is expected that each municipal administration (with a 3-year term) will initiate new plans every 3 years. In many U.S. border cities, urban zoning or transportation projects are considered plans and the policy rationale frequently refers to availability of money.

Policy Rationale

During 1992 to 1993, I conducted a series of interviews in both cities, asking public officials about their policy rationale to set priorities.[34] Their answers by area are reported below.

Health Services

Ciudad Juárez. Allocation of resources is based on the criteria set by the previous administration.[35] Budget is not assigned by specific areas and is allocated according to specific needs.

El Paso. Most pressing concerns are to increase the number of food inspectors to meet federal guidelines. Water and sewage improvement, especially in the Lower Valley (highest concentration of *colonias*—88.5%), and immunization outreach programs are priorities.

Planning

Ciudad Juárez. Priorities are supported by technical studies. Among the criteria is the highway plan, which supports the city's structure. Design of new transportation routes sets the criteria for paving streets.

El Paso. Priority projects are plans that money is available for and that meet the most objectives. Development of jobs, especially for youth, is a priority as a means of reducing crime. Education needs to be considered to obtain employable youth. Transportation has become a priority because that is where the money is and it creates jobs.

Public Works

Ciudad Juárez. Priorities are set on the basis of the city's master plan for urban development. The department also bases its decisions on requests from people. Street paving is decided by preliminary studies of main streets, the need to improve public transportation, and the restructuring of conflicting intersections because of traffic problems or flooding. Paving of *colonias* is based on requests from people. Another interviewee said that when it comes to public works for the community, the planning department intervenes to determine which jobs have priority and in what part of the city.

El Paso. The transportation department's mobilization priority is safe and free flow of people and goods through the city. Air quality is a priority because the Environmental Protection Agency (EPA) now has to approve every single project. No project will be approved unless the department of transportation can prove that it will improve air quality.

Social Development

Ciudad Juárez. The office operates on the basis of the *Solidaridad*[36] programs. The budget is based on those of previous administrations and responses to social demands.

El Paso. According to the chief financial officer, priorities are fairly standard for every city: fire stations within 3 miles of each house, general welfare, water, utilities, transportation, and so forth. Top priority is law enforcement in the broadest sense, including education and prevention. The second priority is economic development, which does not require a great deal of money and is a slow, long-range process. Third priorities are public libraries and transportation. The public priority is "No more taxes." The city must determine if this means that potholes and present quality of libraries are all right as they are.

Water Authority

Ciudad Juárez. Priorities are determined by demographic growth. Eight percent of the budget is used to provide water and sewage in the periphery and 16% for repairs.

El Paso. The public service bureau's priority is to provide safe and adequate supply of water within the city limits and to ensure adequate supply of water in the future. The quality and quantity of water available is a priority. Public concerns are determined by letters and meetings open to the public, although the purchase of surface water rights is limited because of political constraints.

Urban Planning

In Ciudad Juárez, the departments of planning, public works, and social development guide their budgets by the federal allocation through the *Solidaridad* program. Once the money is allocated, state and municipal governments match the funds for specific projects. These departments also raise funds with the local treasurer and have loans from the World Bank and BANOBRAS (the Mexican development bank that funds public works at the municipal level).

Different political parties are expected to have distinct social agendas. This is why it is surprising to see that some of the basic issues had no policy rationale. Apparently, municipal governments continue an inertia set by social reality, and political parties do not think they can transform this reality.

In Ciudad Juárez, the director of planning was aware of the need for comprehensive planning to create a model city. This included reorganizing public transportation and density. At the same time, he wanted to stop and regulate urban growth and coordinate the different providers of public services.[37]

Lack of coordination is one of the causes for deficient provision of public services. One example is a governmental dispute over land reserves[38] to stop irregular settlements during 1993. The planning department thought that ideally land was needed to the east, parallel to the Rio Grande, or in the southeast part of the city, but the secretary general of the city government was negotiating land to the south, where there was no water. The planning department said that the new development master plan tried to avoid growth to the south (Arroyo, 1993).

Texas has no mandatory land use plan. The public service board regulates urban growth, buying and selling land for profit with no guidelines on how to induce urban growth. An array of institutions deals with urban development but apparently with little coordination between them.

Disorganized urban growth is common to the area. One extreme case is that of the *colonias*. The U.S. portion of the Paso del Norte region contains 122 *colonias*, 9,522 dwellings, 17,742 lots, and an estimated population of 47,827 (Texas Water Development Board, 1989).[39] Claims are made that up to 80,000 residents occupy housing in these areas (RGCOG, 1993, p. 4).

One of the explanations for the lack of services is that *colonias* are located outside the city limits and the county has no resources. But perhaps it is the lack of political interest to serve all these people. The most pressing question is why the county does not enforce regulations.[40] Public transportation, for example, is nonexistent outside the city limits, and the municipal transportation company (Sun Metro) is in fact restricted by the state of Texas from establishing routes in the county unless commissioned by the county to do so.

Despite the lack of official number of *colonias* in Ciudad Juárez, city officials recognize informally the existence of 300 *colonias.* Anapra, an irregular settlement with numerous *colonias* bordering the Rio Grande to the west of the city, houses approximately 5,000 families.[41] According to Arreola and Curtis (1993) "shantytown development accounts for three-fifths of the city's land area" (p. 40).

In Mexico, *colonias* are part of a clientele system and are often subjects of political trading between political bosses and governmental officials. They also give rise to corruption whereby the *colonia* leader sells protection and/or promotion of legal title.[42] The phenomenon is so widespread that different ideological groups follow the same track of political negotiation; thus, the possibility of controlling urban growth is remote and complicated. In 1993, the mayor of Ciudad Juárez expropriated 1,200 hectares, enough to solve the deficit of 47,000 houses. Still, people who live in creekbeds or on hillsides refuse to move to the new settlements and pay for the land that includes services, although living in irregular settlements also costs them money and they get no title.

Planning a Border Metropolis

Planning in border cities is incipient with a conspicuous absence of quality strategic planning.[43] Many times, reports for problems are produced when the problems have already changed: Consequently, the conclusions or the projects are obsolete (Graizbord, 1986). Planners risk wasting their energies working on solutions to immediate problems without using a systemic approach to the city as a whole.

Transborder planning is not a new idea. It has been devised as an instrument for the optimal use of natural resources and efficient land use allocation and for avoiding the duplication of services (Graizbord, 1986). But in the mind of public officials, the need for bi-national planning is not yet a priority. If bi-national metropolises on the U.S.-Mexican border do not start planning as a unified region, however, they will hardly become competitive.

The awareness for the need of transborder planning must be based on the idea that decisions cannot be made without considering the neighboring jurisdiction and that the political division between countries cannot stop pollution, water, air, human, and trade flows. The latter happens with or without governmental permits, creating problems that conventional solutions cannot solve.

Planning with multijurisdictional authorities is no simple matter. Some metropolitan areas have overcome this hurdle, but for some others, proximity and common resources have been a matter of litigation and dispute.[44] The problem relates to political constituents and sovereignty.[45] For the U.S.-Mexican border area, exploring the idea of bi-national planning is especially complicated because the territory encompasses 2 countries, 3 states, 5 counties,[46] and 11 cities.

The political boundaries inhibit border regional development, setting a geographic limit for all sorts of activities. The border population must devise a way to avoid political constraints from the three levels of government. "We hope that officials and citizens will begin to see the division, to understand that it is critical, and that it is urgent that they attend to it" (K. Lynch & D. Appleyard, quoted by Herzog, 1986, p. 3).

It is not a cliché to repeat that the solution to border problems must be produced at the border. Solutions for present problems must be supported and funded by both federal governments, following the criteria of the local population. Border people know what their problems are and can devise the solutions they would like to apply to the function of the area. Because regional planning goes beyond local needs, border planning must have input from the state and federal governments. Because of the border's function, studies and projects to ameliorate problems must be conducted jointly by the three levels of government from both countries.

This planning idea has to be based on a democratic rationale. In a democratic system, participation is a fundamental component for sound policy making and planning. There is no reason why the border population cannot participate with all levels of government in deciding on its future.

Planning is a complex, interactive, and dynamic process (Dror, 1990). When planning involves two countries, its complexity increases, requiring basic and fundamental redefinition, for example, of sovereignty. It also requires the elimination of bias and prejudice. Many economic, political, administrative, legal, and cultural differences that include national pride, biases,[47] and prejudice from both parts deter regional planning.

Policy Needs

Although the city has been interpreted as an economic system, it can also be interpreted as a cultural system (Arreola & Curtis, 1993). I suggest that in the Mexican case, it can also be interpreted as a political system because of the widespread use of political criteria in land use and growth management. Also, to eliminate the antigovernment impact of political demonstrations, which are an important aspect of Mexican political life, the government decided to control the public space (Beezley, 1987). One example is the concentration of economic and political power in Mexico City and the accommodation of economic activities in exchange for political support. The extreme case is that of street vendors in downtown Mexico City being granted permission to sell in exchange for support for the ruling PRI.

The Mexican city is an instrument for political control, and its objective is to generate political stability. Because the opposition party, *Partido Acción Nacional* (PAN), won its first victories in Baja California and Chihuahua—both border states with strong urban concentrations—the federal government considers border cities as political risks. PAN's victories are a good sign of democratic development and proof that the city can play a new political role and be reconverted; thus, it might force the PRI federal government to rethink its urban agenda.

Mexico, in general, and the border cities, in particular, must redefine the urban space. The city must be a space for economic, social, and educational opportunities. The city must be used with the main objective of improving quality of life and consequently increase its economic potential. The new United States-Mexico integration process to be stimulated by NAFTA requires a new developmental approach, a new focus on regional interactions, and, of course, a redefinition of the border bi-national metropolis.

Globalization is moving both countries into a macroadjustment, in which regions must start playing an increasing role. The regions must approach the best way to use their potential to improve their international integration.

Regional integration will require fiscal coordination, the development of environmental brown and green agendas focusing on a new developmental philosophy, and a new concept of progress based on sustainable development. The solution of immediate problems will continue consuming most of the cities' and regions' energies; with creativity, however, border regions will complement each other and play an important role for both economies. The bi-national metropolis will be able to compete in the international scene, using their respective national advantages or creating new ones. This, of course, will have important and unforeseen impacts on the urban scene.

Urban development needs to be considered not only as a housing problem but also as a question of structure of services, including the environment, water (drinking and sewage), toxic and solid waste disposal, supplies (supermarket, grocery, and other items), education, and mail. Other bi-national urban considerations include disaster preparedness, law enforcement (because criminals can move back and forth), pollution generated in one city but affecting its neighbor, and garbage generated in one city and disposed in the other.

Education for strategic planning is crucial. City planners and the people in both countries must start to perceive the area as an integrated region, creating awareness that solutions for present problems and prevention for future problems must be bi-national and systemic. Generating a bi-national group of professional planners with the skills to communicate between themselves will be one of the main future challenges.

Conclusion

The U.S.-Mexican border region is stricken by poverty and economic disadvantage. Its limited infrastructure is insufficient to serve over 2,000 *maquiladora* plants and to increase trade and development. In the El Paso-Ciudad Juárez region alone, border crossings amount to more than 3 million people a month. Pressures on infrastructure are expected to grow when NAFTA goes into full effect.

The U.S.-Mexican border region faces a dilemma. It can follow the traditional industrialization model, the one that created environmental, social, and economic problems in many parts of the world. Or it can search for a new model of sustainable development, creating a green and a brown agenda. The decision must be based on strategic planning.

Bi-national planning is needed to face the many challenges. Bi-national planning will advance the region's competitiveness and facilitate the attraction of businesses that can see an advantage in moving to a large market with international linkages. A new, integrated scale in the agglomeration economy and the search for new production ideas will convince investment to move to take advantage not of cheap labor but of an expanded market, international opportunities, and a better developed and competitive border region.

Greater planning capabilities and a developed bi-national culture will help formulate joint scenarios and a more efficient use of scarce resources.

Effective planning will make more resources available to such a degree that an exponential development could be expected.

The literature agrees that planning is the solution for a better development. One idea worth exploring to support a new urban border development is the creation of a bi-national urban planning institute to develop regional strategic planning via a metadefinition of space, focusing on territorial organization and integration of both societies. This institute could help go beyond administrative terms and create a planning memory. Many plans disappear or lose validity without being revised by city planners. Creating a planning history would help institutional reinforcement and planning skills.

Producing planning information in a format useful for future development is a priority to develop mechanisms for the coordination of all levels of government from both countries. It has no less importance in analyzing and overcoming bureaucratic, political, and cultural impediments to urban development.

A bi-national institute would facilitate communication between city planners and policymakers on both sides of the border. It could gather data, analyze information, provide reports for policymakers, conduct special studies, and generate alternative bi-national plans.

The institute must be autonomous and at the same time have backing and funding from both governments, multinational institutions such as the World Bank, and private funding. To avoid the consequences of political change, the institute would need to create a strong base of social support, which could be done through a bi-national advisory board including city, state, and federal planners; professional groups; academics; nongovernmental organizations; community, business, and labor leaders; and economic development institutions.

At first the institute could focus on a specific border subregion, such as the Paso del Norte, but it must be conceived as a borderwide initiative. Interaction with bi-national academic institutions would be valuable because networks of scholars could easily organize the academic input. Identification of urban studies experts in border universities and expert planners working in different corporations is crucial for the success of this new planning experience. The institute could interact with policy institutes and urban studies centers to effectively coordinate advanced training and education for policy specialists and policymakers.

The border region poses many challenges. The most important, perhaps, is to create national awareness of its importance and value for both nations. The border must be perceived as the beginning of the country and the development point of bi-national friendship and understanding. Its cultural and social value might surpass its economic role, which requires even more attention and expanded contribution from federal and state governments from both nations. The border needs to change the perception it has of itself and others have about it. Border people need to improve their regional self-esteem to be perceived as able to cope successfully with different challenges.

North America is moving toward greater integration. The border region is a testing site for the advantages and hurdles to integrate the region. To serve as a good development metaphor, the border must put nonconventional ideas such as bi-national urban planning in place. Not only border people but also both nations must learn how to take advantage of this living laboratory and use the knowledge produced every day to promote a robust North America.

NOTES

1. Calavita (1993) refers to a study undertaken by the Jacksonville Chamber of Commerce in Florida to develop a model of quality of life including the economic environment, public safety, health, education, the natural environment, mobility, government and politics, social environment, culture, and recreation. A similar project in San Diego included affordable housing and excluded government and politics.

2. The Rio Grande (as it is called in the United States) or Rio Bravo (as it is called in Mexico) marks the U.S.-Mexican border from El Paso to the Gulf of Mexico.

3. Elsewhere I elaborate on this approach (see Schmidt, 1993a).

4. Presidios were garrisons to protect Catholic missions, which had the primary task of converting Indians to Catholicism.

5. Sonnichsen (1943) describes the legendary Texas Judge Bean's move to Mesilla, New Mexico: "The Bean Brothers always headed for a frontier community on the make where pushing Americans could get theirs without too much regulation and interference" (p. 41).

6. Baker (1991) estimated the cost of cleaning up the border at $20 billion.

7. Border people frequently complain that decisions affecting the border are made in Washington, D.C., and in Mexico City, far away from the actual problems.

8. In 1993, the Mexican minimum wage was about $.45 an hour and the *maquila* minimum wage was about $1.00 an hour.

9. Barry and Sims (1993) argue that this is actually the cities' fault. "Many of these cities have for years been trying to sell themselves as havens from high taxes and wages. Cities like El Paso are still actively trying to entice businesses from other states to open maquiladoras in neighboring Juarez" (p. 28).

10. Claims of increasing cases of deformities such as the Mallory children in Matamoros or anencephalic babies in the Lower Rio Grande Valley are a result of toxic materials since the establishment of *maquiladoras* (Barry & Sims, 1993).

11. The Mexican government accounts for *maquiladora* exports as national exports to reduce the trade deficit. These hardly can be considered national exports because assembled

products have only a Mexican value added, but this accounting procedure helps to reduce the trade deficit on paper.

12. This segment of the chapter is based on a previous analysis (see Schmidt, 1993b).

13. Robert Warren, head of statistics for the INS, calculated 1.5 million Mexicans living in the United States and the total number of undocumented people as 3.8 million. According to him, 300,000 new undocumented persons enter the United States every year and 50% of those enter via the U.S.-Mexican border (Mendoza, 1994, p. 9). These figures clearly contradict other INS sources. This matter should be understood in the context of the claim from the California governor that the federal government should pay for the state's expenses resulting from undocumented migrants.

14. In January 1994, the Zapatista Army for National Liberation declared war against the Mexican government. Their demands were based on extreme poverty and the deaths of 15,000 indigenous people every year because of lack of medical care for otherwise curable illnesses. See extensive coverage in the Mexican and international press. For an example of U.S. coverage, see Golden (1994a, 1994b).

15. A discrepancy exists with the Mexican statistics. Hiernaux (1986, p. 60) considers that Tijuana's census information was underestimated and cites the following growth rates: 10% in 1960, 6% in 1970, and 5.2% in 1980. See Table 9.6 to contrast the difference with another source.

16. In some cases, military personnel retire to the border cities in which they served.

17. *Carton* means "cardboard," thus *cartolandia* means "cardboardland."

18. After the 1990 census, a strong dispute started between the Mexican census bureau (see Instituto Nacional de Estadística, Geografía e Informática [INEGI], 1991) and many local authorities who claim that INEGI miscounted people. In 1990, population was 798,499 according to INEGI and about 1,250,000 according to city officials. Nevertheless, even at the local level in Ciudad Juárez, a large discrepancy exists regarding the number of households, ranging from 154,108 households for a population of 685,781 according to the water authority to 204,631 households for a population of 1,140,030 according to the economic development office (Barrera & Castillo, 1993, p. 2).

INEGI (1993) explains the discrepancy on the "legality" of the census, which ignores the floating population that is supposed to be counted in its legal residency. Floating population could be those living in the *colonias.* A national undercounting can be assumed if many families leave their cities and are not counted in the new cities in which they are considered floating population.

19. For an assessment of problems and their solutions in the El Paso-Ciudad Juárez metropolitan area, see Schmidt and Lorey (1994).

20. I thank Jesse Acosta from the City of El Paso planning department for his guidance about poverty in the city and the reasons for its causes. According to him, the methodology for assessing the level of poverty is to calculate 80% of the median family income.

21. Seasonal variations may push the numbers up and down, but it is still above the national average. The September 1993 newsletter for the Center for Latin American Studies, at NMSU, *Frontera Norte/Sur,* reported a 10.1% unemployment rate. In March 1994, the rate decreased to 9.4%.

22. According to Clement (see Chapter 6 in this volume), European border regions suffer certain "handicaps" that are similar to the U.S.-Mexican border region: "lower incomes and higher unemployment rates in their own national context, a peripheral position with respect to national economic and political decision making, and the problems imposed by the propinquity of different legal and administrative systems, poor cross-border communications, and a lack of coordination in public services as well as differences in culture and language."

23. This information was obtained in an interview with Henry King, associate director of the Center for Entrepreneurial Development, Advancement Research and Support, University of Texas, El Paso, June 18, 1993.

24. The previous administrations, which left in 1992 to 1993, arranged for a joint city council meeting. The new mayors barely talk to each other.

25. Among the conditions to enhance international competitiveness, a city requires (a) linkages to the rest of the world; (b) a highly skilled and highly educated labor force; (c) full selection of business and professional services; (d) superior transportation and communication infrastructure; (e) a complex of research institutions; (f) large continentally or internationally engaged firms; (g) an environment conducive to a dynamic set of small- and medium-sized firms; (h) adequate access to venture capital and to finance; (i) attractive industrial sites and office spaces; (j) a supportive regulatory environment; and (k) effective government (see Kresl's Chapter 2 of this volume). Clement (Chapter 6 this volume) adds market-related factors, but he says that when it comes to the final choice of location, qualitative elements such as pleasant climate and a good general living climate appear to be more important than quantitative (cost-related) determinants.

26. At a conference on urban services in El Paso-Ciudad Juárez, one expert pointed out that water is measured with different units of measurement.

27. The project proposes building two loops, an internal loop and an external loop, connecting the new port of entry in Santa Teresa, New Mexico, and west Ciudad Juárez to Zaragoza in the east of the metropolitan region. Ciudad Juárez opposes the opening of the port of entry, fearing population growth in that direction where there is no water. The federal government agreed to open the border to please New Mexico congressional representatives who offered to support NAFTA.

28. My appreciation to Nat Campos, director of planning for the city of El Paso, who provided a copy of this report and who answered various questions from myself and my assistants.

29. See Bartra (1987) and Hall (1973) for discussions about time and space.

30. For a general description of the Mexican fiscal system, see Retchkiman (1980) and Evans (1982).

31. The Mexican environmental agency was never abolished. Brown (1986) may refer to the moment when a governmental reform transformed the Department of Ecology and Urban Development (SEDUE) into SEDESOL and created the Procuraduría Federal de Protección al Ambiente (PROFEPA), the Mexican equivalent of the U.S. EPA.

32. "The two countries' political systems are also very different; in contrast to the U.S., where public agencies are fairly stable organizations, the players in Mexico keep changing" (Kjos, 1986, p. 25).

33. In a meeting at San Diego State University in 1991, a Mexican senator told a group of academics that the analysis of NAFTA made by the Mexican Senate took only 20 days.

34. Public offices were matched to the extent possible. Because El Paso has no office for social development, the priorities set by the city's chief financial officer were used.

35. In 1992, the opposition party (PAN) won the governorship of Chihuahua, the majority in the state legislature, and the majorship in various cities including Ciudad Juárez. The previous administration was controlled by the ruling party (PRI), which has been in power in the country since 1929. It is surprising to see an opposition party ruling under the previous government's priorities.

36. *Solidaridad* is a federal program intended to fight poverty. *Solidaridad* moneys are transferred to the states and cities in addition to the *Convenio Unico de Desarrollo Social.*

Cities and states use *Solidaridad* money for paving streets, remodeling schools, providing scholarships, and other projects.

37. Interview by myself and J. W. Wilkie, November 19, 1992.

38. Territorial reserves in Mexico refer to land owned by the government to be used for public housing. The idea is to make land available to discourage land invasions but not to induce or regulate urban growth in any direction.

39. It is difficult to assess the number of *colonias* because new ones crop up constantly. *Colonias* range in size from 1 dwelling with 45 lots (Mesa Verde) to as many as 640 dwellings with 1,350 lots (Westway).

40. There are rumors about a potential conflict of interest of a county commissioner who apparently owned a large portion of the land transformed into *colonias.*

41. In 1984, 38% of Tijuana's total population, or about 300,000 people, lived in irregular settlements (Hiernaux, 1986, p. 64).

42. When government-owned land is invaded, the government can give possession to the squatters. If land belongs to private owners, the government may expropriate it, pay the owner, and give possession to the squatters. In both cases, it can tax the new owners, but because squatters are always poor people, taxing them can be a source of political conflict.

43. I based my concept of strategic planning on Dror (1990), especially the chapter on a general model of planning (pp. 31-60).

44. An example of such conflict is the long water dispute between Doña Ana County in New Mexico and El Paso, Texas (see Barry & Sims, 1993).

45. I asked Fernando Solana, former Mexican secretary of foreign affairs, what he thought would be Mexico's position on creating a bi-national authority. His adamant response: No, for sovereignty reasons.

46. Here I equate the county to the Mexican *municipio.* In this case, two counties and three *municipios* are included.

47. In the fall of 1991, I participated in a conference with the purpose of creating a bi-national water institute. After the first day, one of the Mexican participants asked me: "What is that they really want, why they treat us so nice." For Mexicans, the fear of a hidden agenda is present. For the United States, the prevailing idea is that they deal with a Third World country that is unfairly going to take advantage of them.

REFERENCES

Arreola, D. D., & Curtis, J. R. (1993). *The Mexican border cities: Landscape anatomy and place personality.* Tucson: University of Arizona Press.

Arroyo, J. (1993, May 5). Los terrenos del Lote Bravo no entran en proyectos de planeación [Land in the Bravo lot are not considered in planning projects]. *El Norte,* p. 1A.

Baker, G. (1991). Mexican labor is not cheap. *Río Bravo, 1*(1), 7-26.

Balboa Reyna, F. de M. (1991, August 30). Antes de que la verdad interrumpa . . . [Before truth interrupts]. *La Jornada Laboral,* pp. 4-5.

Barrera, E., & Castillo, L. (1993). *Vivienda y autogestión* [Housing and self-government; mimeo].

Barry, T., & Sims, B. (1993). *The challenge of cross-border environmentalism: The U.S.-Mexico case.* Albuquerque, NM: Resource Center Press.

Bartra, R. (1987). *La jaula de la melancolía* [The cage of melancholy]. Mexico City: Grijalbo.

Beezley, W. H. (1987). *Judas at the jockey club.* Lincoln: University of Nebraska Press.

Brown, H. (1986). Air pollution and problems in the Tijuana-San Diego Air Basin. In L. A. Herzog (Ed.), *Planning the international border metropolis* (Monograph Series, No. 19, pp. 39-44). San Diego: University of California, Center for U.S.-Mexican Studies.

Calavita, N. (1993). Measuring "quality of life" in San Diego. In N. C. Clement & E. Zepeda (Eds.), *San Diego-Tijuana in transition: A regional analysis* (pp. 17-30). San Diego, CA: San Diego State University, Institute for Regional Studies of the Californias.

Center for Latin American Studies. (1993, September). [Table of unemployment statistics]. *Frontera Norte/Sur, 1*(3), 8. (Newsletter available from Center for Latin American Studies, New Mexico State University, Las Cruces)

Clement, N. C., & Zepeda, E. (Eds.). (1993). *San Diego-Tijuana in transition: A regional analysis.* San Diego, CA: San Diego State University, Institute for Regional Studies of the Californias.

Diez Canedo, J. (1984). *La migración indocumentada de México a los Estados Unidos* [Undocumented Mexican migration to the United States]. Mexico City: Fondo de Cultura Económica.

Dillinger, W. et al. (1992). *Juárez: Urban issues survey* (INURD Working Paper No. 16). Washington, DC: World Bank, Urban Development Division.

Dror, Y. (1990). *Enfrentando el futuro* [Facing the future]. Mexico City: Fondo de Cultura Económica.

Evans, J. S. (1982). *The evolution of the Mexican tax system since 1970* (Tech. Papers Series, No. 34). Austin: University of Texas, Institute for Latin American Studies, Office for Public Sector Studies.

García Amaral, M. L. (1993). *Del programa de la estructura urbana del país* [On the country's urban structure program]. Ciudad Juárez: Universidad Autónoma de Ciudad Juárez, Unidad de Estudios Regionales.

Godwin, K. (1977). Mexican population policy: Problems posed by participatory demography in a paternalistic political system. In L. Koslow (Ed.), *The future of Mexico* (pp. 145-168). Tempe: Arizona State University, Center for Latin American Studies.

Golden, T. (1994a, January 3). Mexican troops battling rebels; toll at least 56. *New York Times,* pp. A1-A5.

Golden, T. (1994b, January 4). Rebels determined to "build socialism" in Mexico. *New York Times,* p. A1.

González, V. (1993, July 7). Generan los Mexicanos 20% de empleos en EP [Mexicans generate 20% of jobs in El Paso]. *El Norte,* p. 1A.

Goodman Corporation. (1992). *El Paso-Juárez mobility improvement program* [Mimeo]. El Paso, TX: Author.

Graizbord, C. (1986). Trans-boundary land-use planning: A Mexican perspective. In L. A. Herzog (Ed.), *Planning the international border metropolis* (Monograph Series, No. 19, pp. 13-20). San Diego: University of California, Center for U.S.-Mexican Studies.

Griswold del Castillo, R. (1990). *The Treaty of Guadalupe Hidalgo: A legacy of conflict.* Norman: University of Oklahoma Press.

Hall, E. T. (1973). *The silent language.* New York: Doubleday.

Herrera-Sobek, M. (1979). *The bracero experience: Elitelore versus folklore.* Los Angeles: UCLA Latin American Center Publications.

Herzog, L. A. (Ed.). (1986). *Planning the international border metropolis* (Monograph Series, No. 19). San Diego: University of California, Center for U.S.-Mexican Studies.

Hiernaux, D. (1986). *Urbanización y autoconstrucción de vivienda en Tijuana* [Urbanization and housing self-construction in Tijuana]. Mexico City: Centro de Ecodesarrollo.

Instituto Nacional de Estadística, Geografía e Informática (INEGI). (1991). *XI censo general de población y vivienda, 1990* [Eleventh general census of population and housing, 1990]. Mexico City: Author.

Instituto Nacional de Estadística, Geografía e Informática (INEGI). (1993). *Elementos para la evaluación de los datos censales sobre los volumenes de población y viviendas en el municipio de Juárez, Chih.* [Elements for the evaluation of census data on the volume of population and housing in the county of Juárez, Chihuahua]. Aguascalientes, Mexico: Author.

Inter-Hemispheric Education Resource Center. (1993). *NAFTA-related border funding: Separating hype from help.* Albuquerque, NM: Resource Center Press.

Kjos, K. (1986). Trans-boundary land-use planning: A U.S. perspective. In L. A. Herzog (Ed.), *Planning the international border metropolis* (Monograph Series, No. 19, pp. 21-26). San Diego: University of California, Center for U.S.-Mexican Studies.

Küsel, C. (1988). Tijuana: ¿Una ciudad donde fluyen leche y miel? [Tijuana: A city where milk and honey flows?]. In V. Klagsbrunn (Ed.). *Tijuana, cambio social y migración* (pp. 11-48). Tijuana, Mexico: Colegio de la Frontera Norte.

Lewis, O. (1961). *The children of Sanchez.* New York: Random House.

Lomnitz, L. (1987). *Como sobreviven los marginados* [How the marginal population survives]. Mexico City: Siglo XXI.

Lorey, D. (1993). *United States-Mexico border statistics since 1900: 1990 update.* Los Angeles: UCLA Latin American Center Publications.

Luiselli, C. (1980). *El Sistema Alimentario Mexicano (SAM): Elementos de un programa de producción acelerada de alimentos básicos en México* [The Mexican food program: Elements of a program of accelerated production of basic foods in Mexico] (Working Papers in U.S.-Mexican Studies, No. 22). San Diego: University of California.

Mendoza, F. (1994, May 10). Radican ilegalmente en Estados Unidos 1.5 millones de mexicanos: Robert Warren [1.5 million Mexicans live illegally in the United States: Robert Warren]. *Uno mas Uno,* p. 9.

Realizarán México, EU y Canadá su primera cumbre trilateral sobre transporte regional [Mexico, the United States, and Canada will have their first trilateral summit on regional transportation]. (1994, January 12). *Uno mas Uno,* p. 17.

Retchkiman, B. (1980). *Política fiscal mexicana* [Mexican fiscal policy]. Mexico City: UNAM.

Rio Grande Council of Governments (RGCOG). (1993). *A preliminary evaluation of environmental risk factors for colonias in the El Paso/Ciudad Juárez area* [Mimeo]. El Paso, TX: Author.

Rotella, S. (1993, December 20). Tijuana mayor mounts assault on flood danger. *Los Angeles Times,* pp. A1, A32-A34.

Sainz, P. (1993). *Informe sobre la magnitud y evolución de la pobreza en México en el período 1984-1992* [Information on the size and evolution of poverty in Mexico in the period 1984-1992; mimeo]. Mexico City: CEPAL/INEGI.

Santibañez, J., Valenzuela, J., & Velazco, L. (1993, May). *Migrantes devueltos por la patrulla fronteriza* [Migrants returned by the border patrol]. Paper presented at the conference on Facets of Border Violence, El Paso, TX.

Schmidt, S. (1993a, October 31). La concepción de la frontera [The concept of border]. *Uno mas Uno,* p. 1.

Schmidt, S. (1993b). Migración o refugio económico: El caso mexicano [Migration or economic refugees: The Mexican case]. *Nueva Sociedad, 127,* 136-147.

Schmidt, S., Gil, J., & Castro, J. (1994). *El desarrollo urbano en la frontera México-Estados Unidos: Estudio Delphi en ocho ciudades fronterizas* [Urban development in the Mexico-United States border: Delphi study in eight border cities]. Albuquerque, NM: XII Association of Borderland Studies.

Schmidt, S., & Lorey, D. (1994). *Policy recommendations for managing the El Paso-Ciudad Juárez metropolitan area* (PROFMEX Urban Studies Series). El Paso: University of Texas, El Paso Community Foundation and Center for Inter-American and Border Studies.

Secretaría de Hacienda y Crédito Público. (1980). *Disposiciones legales aplicables a las zonas libres y franjas fronterizas del país* [Laws applicable to free zones and border regions in the country]. Mexico City: Author.

Smith, C. (1992). *The disappearing border.* Stanford, CA: Stanford Alumni Association.

Sonnichsen, C. L. (1943). *Roy Bean: Law east of the Pecos.* Lincoln: University of Nebraska Press.

Stoddard, E. R. (1992). Texas higher education and border funding inequities: Implications for border universities and transborder cooperation. *Río Bravo, 2*(1), 54-75.

Texas Employment Commission, Economic Research and Analysis Department. (1990). *Labor force estimates annual averages.* Austin: Author.

Texas Water Development Board. (1989). *Water and waste water management plan for El Paso County, Texas.* Austin: Author.

U.S. Department of Justice. (1993). *INS fact book: Summary of recent immigration data.* Washington, DC: Government Printing Office.

U.S. Department of Labor. (1991). *Employment and earnings* (Vol. 38). Washington, DC: Bureau of Labor Statistics.

Vellinga, M. (1993). *Industrialización, burguesía y clase obrera en México* [Industrialization, bourgeoisie, and working class in Mexico]. Mexico City: Siglo XXI.

Vernez, G., & Ronfeldt, D. (1991). The current situation in Mexican inmigration. *Science, 251,* 1189.

World Bank. (1991). *Urban policy and economic development: An agenda for the 1990s.* Washington, DC: Author.

10 Achieving Sustainability in Cascadia: An Emerging Model of Urban Growth Management in the Vancouver-Seattle-Portland Corridor

ALAN F. J. ARTIBISE

Introduction

In recent years, the concept of closer cooperation within the Cascadia region (see Figure 10.1) has become increasingly popular. As nations have responded to the restructuring of the global economy, natural regional alliances have been stimulated. In a North American context, for example, the Pacific Northwest-Alaska is a small player. If that regional market is expanded to include British Columbia and Alberta, however, it then ranks as one of the largest in North America. On an international scale, the same principle applies. The two nations and the two regions can bring complementary strengths to the international marketplace.

Several Cascadia regional organizations and activities have developed in response to a perceived commonality of interests across borders. These organizations include the following:

- The Pacific Northwest Economic Region (PNWER), a regional association whose primary membership is state and provincial governments (at the political and officials levels) from Alaska, British Columbia, Alberta, Washington, Oregon, Idaho, and Montana. With staff based in Seattle, PNWER provides a governmental vehicle for regional economic cooperation.
- The Pacific Corridor Enterprise Council (PACE), a private sector regional organization with chapters in Oregon, Washington, British Columbia, and Alaska, with staff based in Seattle. PACE was formed to encourage closer business, trade, and tourism links throughout the region.

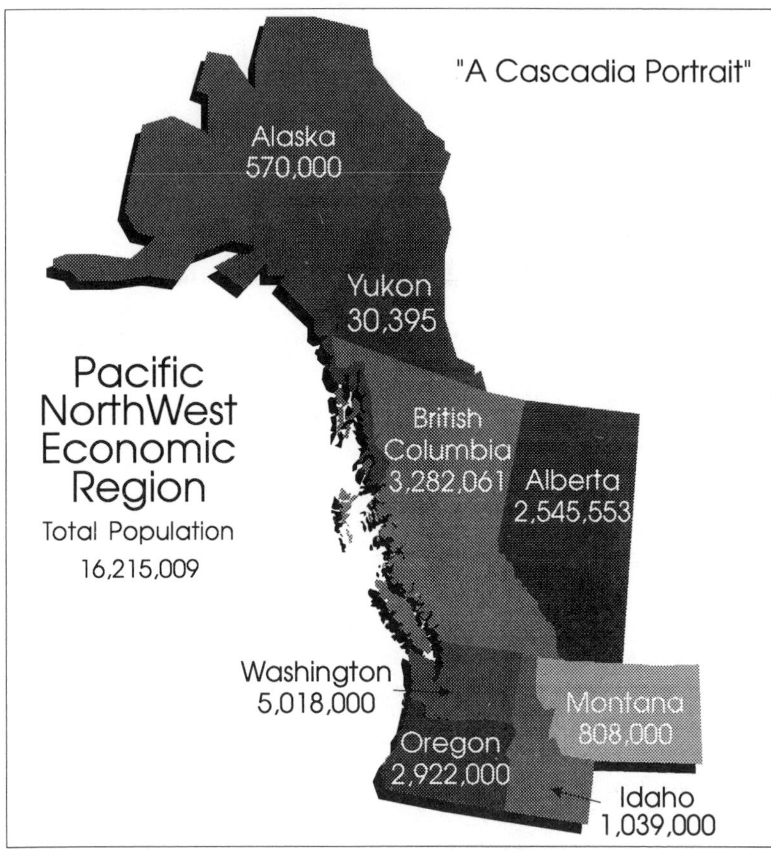

Figure 10.1. The Cascadia Context

- The Cascadia Transportation/Trade Task Force, a strategic alliance along the urbanizing I-5 corridor from Eugene, Oregon, to Vancouver, British Columbia, staffed by the Discovery Institute of Seattle. The task force represents a coalition of all levels of government and nongovernmental organizations in the three jurisdictions to develop cross-border strategies that focus on growth management, cross-border mobility, and improved regional trade and tourism linkages.
- The International Centre for Sustainable Cities (ICSC), recently established under the Canadian federal government's Green Plan to provide Canadian expertise to designated urbanizing regions in the world. The centre has

selected the Cascadia region as the North American example of a rapidly urbanizing bioregion.

The Georgia Basin-Puget Sound Bioregion is an extraordinarily attractive area, and its natural beauty and strategic global location have made it one of the fastest growing urban regions on the continent. Indeed, the region is rapidly merging into a single mega-city with consequent pressures on a common land, air, and marine environment bisected by an international boundary. Given the dramatic growth pressures on the bioregion, special cooperative efforts are needed to preserve the quality of life of the area, including opportunities for employment, housing and recreation. Livable places do not just happen, they are created by the people who live there (Artibise, 1992, p. 8).

Main Themes and Key Issues

The Cascadia region offers a spectacular array of natural and built environments, with wilderness coexisting in relative harmony with sophisticated urban centers. Its geography has few boundaries, and although the international border has produced two different cultures, the citizens of the region have much in common. The region is increasingly attracting attention because of the quality of life and relative prosperity it offers. As a result, in-migration from other parts of North America and overseas is contributing to rapid urban growth.

Although the region, when viewed at night from a satellite, appears as one entity, the jurisdictional lines that crisscross it impose boundaries that inhibit cross-border collaboration. All levels of government in both countries are confronting similar public policy challenges of rapid growth and urbanization, but there is an absence of an established bi-national framework for regional cooperation.

As leaders throughout the region increasingly interact across the border, their common interests become apparent and alliances begin to form. The main issues that drive this cooperation relate to sustainability and rapid urban growth, transportation, trade, tourism, and economic development.

Sustainability

In its *Georgia Basin Initiative* (1993), the British Columbia Round Table on the Environment and the Economy (B.C. Round Table) defined *sustainability* as "ensuring that our use of resources and the environment today does not damage the prospect of their use by future generations" (p. 14).[1] The B.C. Round Table outlined the principles for the foundation of sustainability as follows:

- Limit our impact on the natural environment to remain within its carrying capacity
- Hold to a minimum the depletion of nonrenewable resources
- Promote long-term economic development that increases benefits from a given stock of resources without drawing down on stocks of environmental assets
- Meet basic needs and aim for a fair distribution of benefits and costs of resource use and environmental protection
- Provide a system of decision making and governance designed to address sustainability
- Promote values that support sustainability
- Preserve and protect the environment (p. 15)

The following sets out some of the key questions facing governments as they incorporate sustainability measures into the decision-making process:

- What is the status of land use and transportation planning throughout the region, and what is the potential for regional harmonization and integration?
- What are the environmental priorities of the region as a whole, and what are the impacts of growth management and resource policies of individual governments on the bioregion?
- What is required to manage growth throughout the region in the short term and to establish mechanisms for managing the doubling of the region's population during the next 20 years?
- What immediate and long-term institutional changes to governance are required to manage rapid growth within each jurisdiction and in the Cascadia region as an entity?

Premier Mike Harcourt issued a press release about the *Georgia Basin Report* on June 21, 1993:

> "Preserving the high quality of life in the Georgia Basin is a government priority. This report [Georgia Basin Report] represents an important contribution to the process of developing a sustainable approach to managing the current rapid growth in the region," said the Premier. He noted the report

was based on extensive consultation with stakeholders in the Basin, including a workshop held in Nanaimo in January this year. (Premier's Office, 1993)

Transportation

The region is served by a complex of transportation facilities (airports, seaports, ferry terminals, highway systems, and railways) and services (international, national, and regional air carriers; cruise ships; local and international ferry services; trucking and motor coach industries; and passenger, freight, and commuter rail services). The rapid growth in population and urban sprawl, coupled with increased tourism and trade volumes, has created stresses on the transportation system throughout the region.[2] This is reflected in all modes, with examples such as serious traffic gridlock in the metropolitan areas, airports exceeding operating capacity, and cross-border delays and congestion.

The following are some of the key questions facing governments and their agencies as they respond to the transportation needs of today and plan for the long term:

1. What are governments doing to better integrate regional transportation requirements with land use planning and to minimize the future environmental impacts and the cost burden of expensive new transportation infrastructure?
2. What new infrastructure investment is required in the region's major ports and airports, and how are these agencies planning to integrate these projects with local and cross-border highway and rail systems?
3. What are the constraints and opportunities affecting the cross-border transportation systems?
4. What is the potential for collective regional action to support new corridor initiatives in transportation, such as high-speed rail and north-south trade?

Trade, Tourism, and Economic Development

Although the traditional large employers in the resource, defense, and aerospace sectors will remain in a dominant position within the regional economy, value-added manufacturing, and knowledge-based, high-technology small and midsized companies are providing much of the recent growth in the economy. Unlike the highly integrated regional economies of the Great Lakes region, built on the automotive and manufacturing sectors, the Pacific Northwest and British Columbia have traditionally been competitors. The ports, airports, and railways serving the region compete fiercely as gateways for international traffic to North America. With parallel resource bases, many of the region's products are similar and competitive in sectors such as forest products, agri-food, wines, and seafood.

With the passage of the Free Trade Agreement (FTA) and the North American Free Trade Agreement (NAFTA), north-south trade links have strengthened and increased. Although much of the trade within the region is resource based, "invisible" trade in the technology sectors is increasingly important, and intraregional tourism is one of the largest sectors of the economy.

The following are some of the questions to be considered by government and business for increasing trade and economic activity within the region and with continental and offshore markets:

- What is the nature and extent of current trade in commodities, goods, and services within the region?
- What sectors offer opportunities for partnerships, supplier contracts, and venture capital investment, and what are the mechanisms that could facilitate their achievement?
- What is the impact of tourism within the region, and what are the trends and opportunities for new cooperative initiatives, both regionally and in offshore markets?
- What is the impact of regulatory and transportation constraints on current cross-border investment, trade, and tourism, and what measures are required to fix these problems and plan for future growth?
- What are the opportunities, both commercial and noncommercial, for cooperation between Canadian and U.S. ports and airports that will enhance the international competitiveness of the region?

"Progress towards a sustainable environment will require changes to current management institutions and practices. . . . There is an urgent need to develop long-term strategies that cut across traditional political, institutional, and community boundaries to create integrated environmental management" (*The Lower Fraser River Basin: A State of the Environment Synopsis,* 1992, p. 9).

Sustainability

Throughout the region, a wide range of initiatives are directed toward achieving the goals of sustainability. These are directed primarily at the region's ecosystems, its growth management strategies, and its institutional arrangements. Many of these activities are multijurisdictional; some are bi-national. Innovative techniques are being employed to find more effective means to meet the tests of sustainability.

Highlights of Sustainability Initiatives

The British Columbia/Washington Environmental Cooperation Agreement, signed by the premier and governor in 1992, established the British Columbia/Washington Environmental Cooperation Council (1993). Led by provincial and state environmental departments, with federal observers, this regional group meets biannually and has established five working groups to address the following priorities:

1. Marine water quality
2. Nooksack River flood management
3. Columbia River/Lake Roosevelt water quality
4. Abbotsford/Sumas aquifer
5. Georgia Basin air quality

In April 1993, the British Columbia premier and the Washington governor issued a joint statement that reaffirmed cooperation on environmental issues and broadened their commitment to regional growth management and transportation issues. It was also agreed that the governor of Oregon would be involved in future discussions, and British Columbia officials have initiated meetings with their Oregon counterparts.

The B.C. Round Table issued its final report (with addendum) in December 1993 on the *Georgia Basin Initiative: Creating a Sustainable Future.* In response to its mandate from the government of British Columbia, the report offers 17 recommendations regarding governance, urban containment, coordinated transportation and energy planning, bi-national economic cooperation, education and public awareness, and government accountability. The report is currently under review by the British Columbia government.

The Greater Vancouver Regional District (1993a, 1993b, 1993c) has recently completed an extensive public consultation process ("Creating Our Future") to develop its *Livable Region Strategy* and the *Transport 2021* long- and medium-range transportation plans. These proposals, currently under consideration by member municipalities, reflect a different approach to strategic planning. For the first time in Greater Vancouver, an effort is being made to plan regional land use and transportation on a metropolitan basis. The strategy describes in physical terms how an additional 1 million residents can be accommodated during the next 30 years while sustaining high levels of environmental quality and livability.

Another example is the Capital Region District of Greater Victoria, which is currently developing its regional transportation strategy and development plan. This plan is based on regional values and goals.

The Fraser Basin Management Program, established in 1992, is a unique 5-year agreement by federal, provincial, and local governments to chart a course of action to work toward achieving sustainability of the basin. The multiple stakeholders have identified priorities for action, including fish habitat, water quality and quantity, water diversion, liquid and solid waste management, community and regional development, floodplain management, government processes, and program coordination. Strategic programs address the following activities:

- Management strategies in priority areas
- Demonstration projects
- Institutional arrangements
- Indicators for auditing sustainability
- Information collection and management
- Public information and awareness

The Washington State Growth Management Act of 1990 (see De Grove, 1992) requires cities and counties to

- plan for the conservation of important timber, agricultural, mineral and resource lands;
- adopt development regulations for critical areas (wetlands, fish and wildlife habitat, aquifer recharge areas, and flood or geographically hazardous areas);
- establish urban growth boundaries and coordinate their comprehensive planning with neighboring jurisdictions; and
- tie local land use planning to transportation systems that meet "level of service" standards and federal clean air requirements.

Environment 2010 (see Washington State, 1989) is a joint project of the U.S. Environmental Protection Agency and the state of Washington.

It has an action agenda that addresses air and water quality, habitat preservation, protection of agricultural and recreational lands, energy and water conservation, and public education.

The Washington Trade and Economic Development Department's Sustainable Development Task Force is based on the premise that it is necessary to remove the idea of a conflict between jobs and the environment. Accordingly, it has a mandate to establish a set of principles for guiding economic development in the context of sustainability. Its five work groups concentrate on energy, natural resources, sustainable communities, water and recovered resources, and education-public awareness.

Oregon has had land use legislation in place since the passage of Senate Bill 100 in 1973 (see Abbott, Howe, & Adler, 1994), and all county and local plans were adopted by 1986. A public interest group, 1000 Friends of Oregon, has been effective in withstanding challenges and strengthening the legislation. A recent study (Ketcham & Siegel, 1991) found that

- housing is more affordable in Portland than in any other major West Coast city,
- the average size of an undeveloped lot had dropped from 13,000 square feet to less than 9,000 square feet between 1979 and 1983, and
- the amount of land zoned for multifamily housing had jumped from 7% to 28%.

Oregon Benchmarks (see Oregon State, 1991) were developed by six steering committees that were formed to propose benchmarks important to Oregon's progress. The benchmarks were submitted to the legislature in 1991 as measures of the state's well-being:

1. People: adult health and literacy, drug-free teens, workforce adaptability, and job skills preparation
2. Quality of life: air quality, affordable housing, health care access, urban mobility, and crime prevention
3. Economy: personal income, economic diversity, manufacturing for export, public infrastructure investment

The Oregon Transportation and Growth Management Program offers joint federal-state funding to local governments to respond to the pressures of urbanization and rapid growth. The funding is to be used

- to update transportation plans to enable more people to make some of their trips by means other than car,

- to examine how changing land use plans could reduce street and highway expenditures and encourage alternative modes of transport to the car,
- to test and demonstrate new tools for urban growth management, and
- to promote pedestrian and transit-friendly development using public information and awards programs.

The following expresses the "Creating Our Future" vision from the Greater Vancouver Regional District's *Livable Region Strategy* (1993a):

> Greater Vancouver can become the first urban region in the world to combine in one place the things to which humanity aspires on a global basis: a place where human activities enhance rather than degrade the natural environment, where the quality of the built environment approaches that of the natural setting, where the diversity of origins and religions is a source of social strength rather than strife, where people control the destiny of their community, and where the basics of food, clothing, shelter, security and useful activities are accessible to all. (p. 1)

Transportation

One of the region's strongest features is the transportation system. Located on the great circle route, midway between Asia and Europe, Cascadia's ports and airports are strategically located as gateways to North America. North-south corridors support trade and travel within the region and with other U.S. states and Mexico.

The Pacific gateways are engaged in vigorous competition for traffic and conduct targeted marketing campaigns to attract shipping lines and airlines to their ports and airports. With this tradition of commercial competition, contact and cooperation between Canadian and U.S. ports have been limited. The north-south corridors offer a range of opportunities for bi-national cooperation. These corridors support economic activity throughout the region. Both sides of the border have mutual interests in enhancing and strengthening these corridors.

> Significant environmental, economic, and social pressures are being exerted on the Georgia Basin-Puget Sound region. In many areas the natural environmental quality of life is deteriorating, primarily as a result of rapid population growth and human settlement patterns, and the over-consumption of natural resources by the region's population. . . . The ability of our air, land and water to absorb the impact of human use has been severely strained,

and cannot be sustained over the longer term. Fundamental changes are required. (B.C. Round Table, 1993, p. 13)

Gateways

The ports and airports of Vancouver, Seattle, Tacoma, and Portland are engaged in a wide range of activities to improve their ability to compete with each other and with California gateways. These initiatives include the following:

- Capital investment in infrastructure, such as container terminals, cruise ship facilities, airport expansion, and dredging programs
- Aggressive marketing programs to attract new carriers, with strong growth in containers, sea/air cargo, grain exports, and cruise trade
- New alliances among ports and between ports and airports within a state or province to provide vehicles for joint marketing opportunities

Some of the competitive issues between the ports of Greater Vancouver and the Pacific Northwest are these:

- The Alaska cruise trade, with more than 500,000 passengers per season, is based in Vancouver, mainly because of restrictive U.S. maritime legislation that prohibits foreign flag carriers (the cruise lines) from operating between two U.S. ports. Seattle is currently supporting legislation in Congress to allow for the carriage of passengers "coastwise" between U.S. ports, a measure that would allow Seattle to compete for the Alaska cruise business.
- Competition for Canadian containerized cargoes is intense. The U.S. West Coast ports and railways have captured more than half of the container volumes imported and exported from Canada; the Port of Vancouver is attempting to regain Canadian cargoes and attract U.S.-bound containers.
- Different taxation systems in the two countries have created significant inequities in the financial resources of competing ports, to the disadvantage of the ports of Greater Vancouver. The U.S. ports are local public entities and receive significant funding from the local tax base; Canadian ports are part of a national system, pay dividends, make large, extraordinary contributions to the federal shareholder, and pay grants in lieu of taxes to the municipalities.

Despite the commercial competition among the ports and airports, several potential areas for cooperation have been identified:

1. The first meeting of the Tri Association was held in New Westminster, British Columbia, in 1993 and included most of the ports from Oregon, Washington, and British Columbia. The agenda addressed operational issues such as electronic data interchange (EDI) and port development; economic issues, such as NAFTA and the forest industry; environmental challenges, such as public process and environmental audits; and Cascadia regional developments, such as the Georgia Basin Initiative and the Cascadia Corridor.
2. The ports in the region could share information and work cooperatively to solve common operational and environmental problems, such as the disposal of dredge material, disposal of ballast water by vessels in port, and the establishment of common agricultural inspection procedures.
3. The ports and airports could offer support for increased government funding for border crossing technology and infrastructure improvements and participate in the planning for intraregional transportation linkages, such as high-speed rail.
4. The ports and airports might consider joint international marketing initiatives in specific targeted markets in which a broader regional approach could potentially attract new business.

Corridors

Several bi-national organizations have formed in recent years in response to the need for regional alliances. Two of these, PACE and the Cascadia Transportation/Trade Task Force, have specifically focused on the need to address the current and future impediments to north-south trade and the efficient movement of goods and passengers throughout the corridors. The impetus for these initiatives has come from traffic gridlock in the metropolitan centers and long delays and congestion at the border crossings. There is a growing awareness that rapid urbanization throughout the region is placing severe stresses on the transportation systems and that bi-national cooperation is essential to attract federal funding and support for the region.

U.S. Initiatives

Rail and Trade Corridor

Two acts of the U.S. Congress, the High Speed Rail Development Act of 1994 and the Intermodal Surface Transportation Efficiency Act (ISTEA) of 1991, have provided a legislative framework for regional responses to some of these north-south corridor issues. The High Speed Rail Development Act designates the Pacific corridor from Portland to Vancouver (Cascadia's "Main Street"; see Figure 10.2) as the only international high-speed rail corridor in the United States. In the short term, high-speed rail service, with Japanese-style bullet train technology, is not planned. There is, however, significant interest in the region in the future of high-speed rail as an essential element of the regional transportation system and as an efficient, environmentally sensitive alternative to highway and air travel. Recently, there have been two high-speed rail demonstration projects in the corridor, the latest of which is the Spanish Talgo high-speed tilt train, which performed very well in a test operation between Seattle and Portland between April and September 1994.

A bi-national coalition will be essential to gain federal authorization and financing for the high-speed rail program. Funding sources, budgetary constraints, and trade-offs with other transportation infrastructure requirements will be major factors in determining the future of high-speed rail in the corridor.

Currently, U.S. government agencies and railways are taking an incremental approach to rail passenger service in the corridor with the reintroduction in October 1994 of conventional Amtrak service between Seattle and Vancouver in just under 4 hours. During the next 5 years, ongoing system upgrading is planned to increase speeds and reduce travel times to be competitive with other modes of transport. The new service has generated public interest along the corridor and has provided a focus for regional cooperation.

The British Columbia/Washington State Rail Working Group (established by the premier and governor in April 1993) is responsible for the bi-national coordination for the new service and has been effectively addressing regulatory and operational issues. In particular, both U.S. and Canadian customs and immigration agencies have agreed to the elimination of border inspections in favor of pre- and postclearances at rail terminals. The group, supported by local officials, is also working with regulatory agencies and communities along the route to gain the approvals necessary for increases to local speed limits.

In the context of the FTA and NAFTA, the ISTEA requires that two studies relating to international trade be undertaken. Recently, the U.S. Department of Transportation (1994) released a report to Congress titled *Assessment of Border Crossings and Transportation Corridors for North American Trade*. The studies were motivated by the history of lengthy

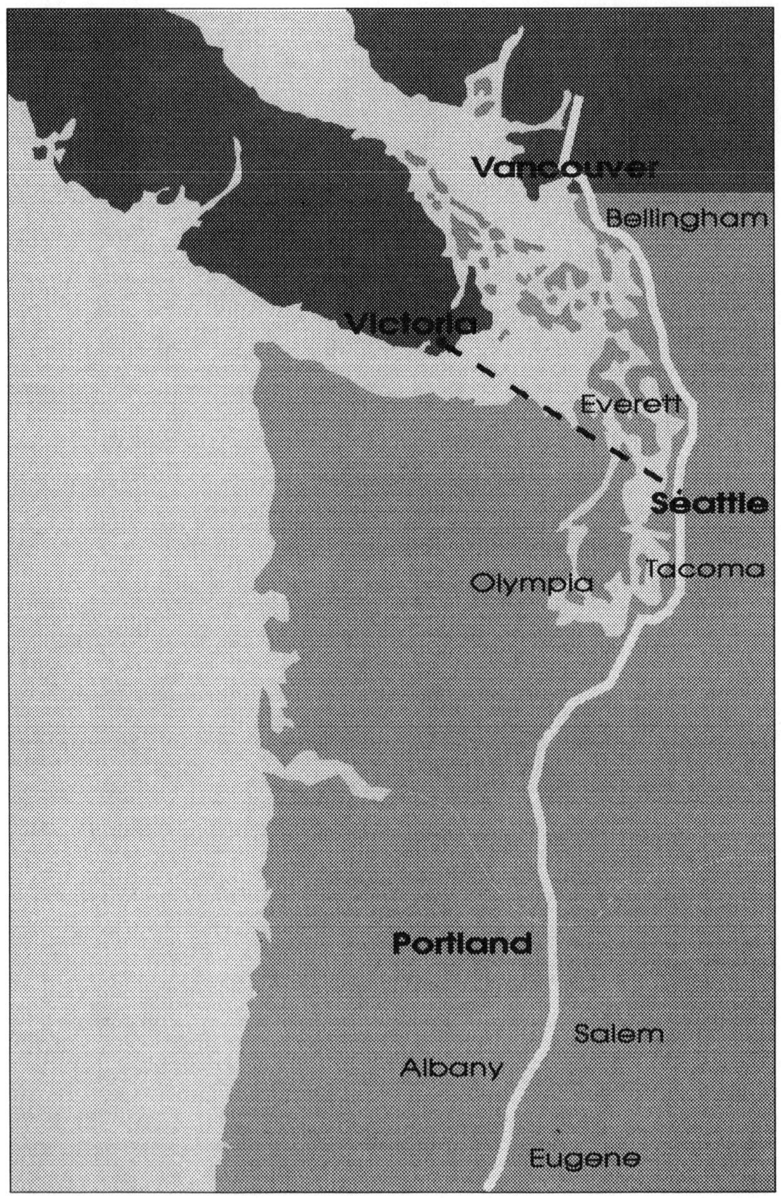

Figure 10.2. Cascadia's Main Street

delays and congestion at international border crossings and the concern that the current and projected increased trade among the three North American nations will outstrip the transportation systems' ability to handle the additional international traffic. Relevant principal findings are the following:

1. Canada is the largest trading partner of the United States, with merchandise trade in 1992 between both countries valued at $189 billion; trade with Canada currently accounts for 20% of total U.S. world trade.
2. Trade between states and provinces in the three eastern regions (Great Lakes, Mid-Atlantic, and New England) accounts for 60% to 65% of exports and imports.
3. In the west, trade flows in three cross-border trading subregions: the Upper Plains, the Rocky Mountains, and the largest, the Pacific Northwest; United States-Canada trade through the western region is projected to increase during the next 10 years—U.S. exports by 16% to 24% and Canadian exports by 24% to 34%.
4. The arterioles providing access to border crossings in the United States are under stress, are badly in need of repair and upgrading, and are unable to handle significantly more traffic.
5. Communities adjacent to busy border crossings face problems of congestion, air pollution, and safety. Border states have not allotted sufficient funds to these communities for border crossing approaches; new funding sources and improved allocation systems are needed.
6. Other causes of crossing delays are complex inspection and cargo clearance procedures, inadequate traffic management, inspection staffing shortages, and lack of coordination of hours of operation.
7. Infrastructure and facilities planning for major border crossings is fragmented and inadequate; planning should include all levels of government and should be bi-national in scope.

Main recommendations are as follows:

- Fully fund the ISTEA to provide additional resources to states for trade-related projects.
- With state and local governments, private financial institutions, carriers and other private interests, develop a range of funding options for infrastructure improvements, emphasizing existing federal, nonfederal, and potential private sources.
- Establish public-private sector task force(s) to address border crossing congestion and to identify and promote the use of new technologies and other non-capital-intensive methods of facilitating the movement of passengers

and goods through major border crossings. Any initiatives should be closely coordinated with the other national governments.

- Provide for a limited number of pilot projects at specific gateways, with funding from a variety of federal, state, local, and private sources.
- The Department of Transportation and other federal agencies should establish bi-national planning zones to ensure coordination for future trade-related infrastructure and technology requirements.

Other Issues

During research for this chapter in the United States, several additional Pacific border crossing issues were identified.

Institutional issues. Integrated bi-national planning is needed among all stakeholders, which include federal agencies, local and regional planners, facility operators, carriers, and shippers. The fragmented border responsibilities of seven different U.S. federal agencies result in uncoordinated planning, inadequate and uneven staffing, and a lack of political clout during federal budget negotiations.

Staffing problems are a challenge at both land crossings and airports; according to the Customs Service, a 10% staff cut is under way in Whatcom County; the Can Am Border Trade Alliance reports that staffing billets in Customs and Immigration are being shifted from the northern tier to the southern tier states, which have benefited from a 5-year Southern Border Improvement Plan.

Technological issues. Through better coordination of agencies and the influence of a cross-border alliance, the new technologies for processing passengers and cargo should be given priority, for example: EDI, truck bar codes, Advanced Passenger Information clearance for air passengers, and Intelligent Vehicle Highway System electronic bulletin board. A bi-national multimodal transportation database for freight and passenger transportation planning is needed.

Infrastructure issues. Currently, an estimated $1 billion annually is lost because of border crossing delays, with 71% of total western United States-Canada trade concentrated at the Whatcom County border crossings. Improvements in access to the border crossings, seaports, airports, and intermodal transfer facilities are critical to support increased trade. Needed also are separated passenger-commercial lanes and longer approaches at border stations. The Cascadia Task Force is working with Washington State Department of Transportation and the congressional delegation to link the north-south corridor (I-5 and Highway 99) to the National Highway System designation in ISTEA to gain eligibility for additional federal funding.

Strong support exists on the U.S. side to expand the PACE lane and improve its marketing. This successful, low-cost system has the potential to significantly reduce congestion. Over 30,000 vehicles are registered under the PACE program and 2,000 vehicles use it daily.

> In many respects [there are] an impressive array of cooperative efforts in the region. . . . Yet, in many ways, the Northwest has made progress where progress came easy, not where it was most needed. . . . Intergovernmental efforts have limited their potential by limiting size and scope. They fail to seek out lessons from others' experiences and apply the best of them across borders and departmental lines. They don't harness existing organizations to help foster new initiatives. . . . There is no shortage of groups interested in economic policy in our states, provinces, cities, counties, reservations, towns, and neighbourhoods. . . . [But] few regional organizations attempt to coordinate the activities of the diverse groups. (*Northwest Resources for Regional Cooperation,* 1990, pp. 21-22)

Canadian Initiatives

Canada Customs

During the past two years, fundamental changes have been under way in Canada Customs and Immigration in its legislative framework, its organizational structure, and the execution of its mandate. Following an extensive consultation process by the Customs Act Review Task Force, with over 175 agencies and organizations, a proposed Canada Customs Act has been drafted to respond to the new realities of Canada's international trade and transportation requirements. The proposed act would provide a more flexible customs regime, with emphasis on voluntary compliance and self-assessment and a shift from a transaction-by-transaction to an exception, audit process (Assistant Deputy Minister A. Coxsedge, Customs Operations, Revenue Canada, personal communication).

A New Business Relationship Initiative has been introduced for the customs commercial process (currently being tested in the aerospace and automotive sectors) that would

- allow importers, exporters, and international carriers to establish a business relationship with Customs on the basis of their needs;
- allow Customs to offer integrated services for the release of goods on behalf of several federal and provincial departments;
- expedite release on the basis of advance information, such as EDI;
- provide for self-assessment of duties and taxes; and
- offer flexible accounting procedures to facilitate client payments.

A shift in government attitudes is also evident, with greater emphasis on facilitation of trade and tourism and less on regulation. Examples of this approach are the "Red Door/Green Door" system for cruise ship passengers in Vancouver, the pre- and postrail station clearances for the new Amtrak service, and the Super Host training program offered by Tourism BC for customs officers.

The PACE program at the Blaine crossing has been successful. Plans are underway to extend this principle to regular importers and exporters under a line release system. The bi-national Border Initiatives Committee, which includes all levels of government and border users, meets annually to discuss common problems and explore new initiatives.

Other Issues

During the research for this chapter in Canada, several additional border crossing issues were identified:

1. Some of the main constraints to commercial traffic are the lack of adequate access and facilities at the Pacific crossing and the mix of auto and truck traffic. Suggestions include the construction of an additional access lane from I-5 to the border, and in the long term, a new commercial border crossing.
2. Other commercial concerns related to staffing shortages, reduced hours of operation by U.S. Customs, and a $25 fee for off-peak hour service by Canadian Customs, which deters truckers from crossing during the night and spreading traffic flows.
3. Auto/tour bus passenger complaints relate to the long border crossing delays that are increasingly common. Two current initiatives could improve this situation—an aggressive bi-national campaign to market the PACE pro-

gram as a cost-effective way to reduce congestion and the introduction of license plate reader technology to reduce inspection times.

The Importance of Common Interests

Kelly (1994) points out the importance of recognizing the varied interests shared by all of Western Canada:

> The opportunity in a larger vision of Cascadia is to redefine our economic horizon and, at the same time, be better able to manage our environmental future. . . . Failure to recognize our common interests reduces us to our traditional positions of Western Canada as a tertiary market of Toronto and, for Northwestern states, as market outposts for Los Angeles, Chicago and New York. Commerce aside, our environment is a shared one. The vision of a larger bioregion—one forest, one waterway, one airshed—should invite us to preserve one of the last great Edens on the planet.

■ Trade, Tourism, and Economic Development

The Regional Economy

The economies of the Cascadia region share many common features. They are highly export oriented with a focus on the Pacific Rim. They have traditionally relied on resource-based industries, which remain the foundation of the region. Increasingly, however, other sectors have been providing most of the growth and employment.

These include resource-based value-added industries, aerospace, manufacturing, defense, transportation, energy, tourism, computer software, environmental industries, and biotechnology. Throughout the region is a complex network of trade relationships and associations, some of which are long-standing in sectors such as energy and the forest industry. Other more recent networks in the technology and service sectors are evolving.

Intraregional trade is significant and growing, with high levels of cross-border commuting, shopping, and movement of goods and services. The predominant commodities traded within the region are newsprint, natural gas, lumber, seafood, and agricultural products. The region is the third largest computer software development center in North America and an expanding center for environmental and biotechnology industries. Tourism is one of the largest sectors of the regional economy, ranking second in British Columbia and fourth in Washington.

Industry Sectors

The current focus of British Columbia's trade efforts with the Pacific Northwest is to encourage the development of the knowledge-based industries. Three sectors with a regional base have been targeted—biotechnology, software technology, and environmental industries. Through government-industry association partnerships, the region is promoted internationally as a technology center.

The biotechnology industry provides a model for regional alliances and synergy in a truly global industry. This relatively young industry, which is engaged in sophisticated scientific research and product development, has an established base of small and midsized firms throughout the region. These firms have complex linkages with partners worldwide. The industry relies for its success on industrial mass, where presence and size are the key. As a region, Cascadia ranks ahead of New York as the sixth largest biotechnology center in North America.

Since 1991, the industry associations in the region have been holding joint meetings, sharing information, and raising profiles. The regional emphasis has attracted the attention of analysts and policymakers outside the region and has given the industry advantages in targeting investment from major North American financial centers. Regionally, the Vancouver Stock Exchange is increasingly an important source of venture capital to the industry.

Economic Initiatives

Some business and government leaders have recognized the potential for expanding Cascadia economic activity, both intraregionally and into third markets. The next sections include examples in the fields of communications, strategic alliances, trade, and tourism.

Communications

The regional publication *The New Pacific,* published in Vancouver, has become an effective vehicle for the exploration of the economic, political, and social issues that affect the Cascadia region, with a focus on the Pacific Rim. A new publication, *Cascadia Forum,* was launched in 1993 at the University of Washington College of Architecture and Planning in joint sponsorship with the University of Oregon School of Architecture and Allied Arts and the University of British Columbia Schools of Community and Regional Planning, Architecture, and Landscape Architecture. The journal will cover the events, ideas, and issues shaping the quality of the built environment in the entire region.

Strategic Alliances

Recently, the Discovery Institute, a Seattle think tank, released its *International Seattle* report (see Hamer & Chapman, 1993). This comprehensive strategic document provides a blueprint for Seattle's role in developing regional alliances as part of a coordinated bilateral advancement of the Cascadia region in the global economy.

With the passage of NAFTA, members of the financial community have proposed the establishment of an expanded international financial center that would draw on the expertise within the region and build on the attributes of two nations, with a focus on the Pacific Rim.

Trade

Joint trade initiatives have been proposed by several organizations throughout the region. Members of PNWER have suggested the establishment of regional trade-tourism offices in third markets such as Japan or Mexico. BC Trade and the Port of Seattle have suggested joint trade missions to market the region. PACE has organized regional meetings, including Mexican business leaders, to explore opportunities for mutual trade and tourism partnerships. The biotechnology associations in the region are planning a conference to highlight their industry in the Pacific Rim. Roseland (1992) has emphasized the importance of using biotechnology to develop communities that are not only economically viable but ecologically sound:

> Sustainable development requires that we develop our communities to be sustainable in global ecological terms. This strategy can be effective not only in preventing a host of environmental and related social disasters, but also in creating healthy, sustainable communities that will be more pleasant and satisfying for their residents than the communities we live in today.
>
> Sustainable communities will emphasize the efficient use of urban space, reduce consumption of material and energy resources, and encourage long-term social and ecological health. They will be cleaner, healthier, and less expensive; they will have greater accessibility and cohesion, and

> they will be more self-reliant in energy, food and economic security than our communities now are. Ecologically sustainable communities will not, therefore, merely "sustain" the quality of urban life—they will improve it. (pp. 339-340)

Tourism

Tourism, industry, and government leaders within the Cascadia region are beginning to recognize the potential for joint regional initiatives in international tourism markets. A recent report for Tourism Canada by the Cascadia Institute (1994), titled *Opportunities for Expanding International Tourism in Cascadia: A Canadian Perspective,* is providing an information base for regional discussions. A Cascadia Tourism Leaders' Forum, with participants from across the region, was held in Vancouver on March 30, 1994. Report findings are summarized as follows:

1. The region offers a wide variety of travel experiences for international visitors, with natural scenery, entertainment and cultural attractions, historical sites, and national parks. Its strongest features include outdoor sports-adventure products, year-round resorts, Alaska cruising, urban experiences, and an extensive, multimodal regional, continental, and international transportation system.
2. During project interviews, the question of broader regional tourism marketing initiatives was explored, both in the context of western Canada and in the Cascadia region. Many of the respondents expressed interest in pursuing more regional opportunities or had suggestions for markets or products that could be developed or expanded with a regional or transborder approach.
3. Despite excellent potential for developing an array of transborder tourism products and marketing jointly overseas, the challenge is to identify specific opportunities and create partnerships in the absence of an established transborder framework.
4. Tourism segments that offer potential in the development of products and overseas markets for joint Cascadia regional initiatives are the Alaska cruise trade, northern partnerships (Alaska, the Yukon, and British Columbia), regional rail touring, the Seattle/British Columbia Golden Triangle (Vancouver, Whistler, Victoria), aboriginal tourism, a pilot project in the Association of Southeast Asian Nations (ASEAN) region, and a Cascadia intermodal travel pass.

5. It is recommended that this Cascadia tourism initiative be coordinated with the activities and programs of regional organizations, in particular, the Cascadia Transportation/Trade Task Force, ICSC, and PNWER.

6. It is recommended that a bi-national regional tourism planning meeting be held in Vancouver, just prior to the task force meeting, on March 30, 1994. Invited participants would be representative of the private and public tourism sectors across the region. The primary purpose of the meeting would be to consider potential opportunities for Cascadia regional tourism initiatives, to develop an implementation plan and assign responsibilities for its execution. A report from this meeting would be presented for information at the task force meeting the following day.

The PNWER Tourism Working Group (consisting of 26 legislators, public program administrators, researchers, and private citizens) commissioned a California Tourism Study in 1992 to identify tourism and travel plans. The regions' tourism agencies, together with the U.S. Forest Service and the California Office of Tourism, jointly funded the market research survey in six top California markets. One of the principle findings was that California's image of the Pacific Northwest is limited to Washington and Oregon and that British Columbia is not perceived as part of the regional destination. The report, completed in 1993, will be used by the participants in their individual marketing programs; joint efforts will also be explored (see Cascadia Institute, 1994).

■ Cascadia Opportunities

In doing research for this chapter, I found that people throughout the region are searching for new approaches to the management of their environments. Broad recognition exists that the traditional structures of government are no longer effective in responding to the challenges of an increasingly interdependent world. Examples abound of new and innovative approaches to planning and organization that cut across jurisdictional boundaries and have multiparty stakeholders and a broad regional mandate.

These multiparty forums produce recommendations, plans, agreements, and protocols, which typically fall back for decisions and delivery on the traditional, accountable levels of government. Because governments are hierarchical and compartmentalized, they have difficulty in responding to forum initiatives, especially when a clear political direction is lacking.

New models for planning and governance are needed. For example, the B.C. Round Table, in its *Georgia Basin Initiative,* offers a range of new institutional options that recognize the geographic realities of the bioregion, the need for specific political accountabilities, and regional organizational mechanisms of governance. The report calls on the British Columbia government to give urgent attention to this issue.

Questions of governance are highly complex within each jurisdiction and are compounded when larger regional alliances across international boundaries are formed. Success depends on how effectively leaders on both sides can overcome differences in culture, political systems, and organizational structures to create a framework for bi-national cooperation to meet specific objectives.

In the Cascadia region, the bilateral agreements between British Columbia and Washington have resulted in specific environmental, economic, and transportation projects. Multiparty public and private sector regional organizations have been formed to respond to regional economic, trade, and transportation issues. These include PACE, PNWER, and the Cascadia Transportation/Trade Task Force.

These governments and organizations have been pioneers in breaking down some of the natural barriers created by the international border to address specific regional needs. They have demonstrated results and their experience is instructive in exploring other opportunities for bi-national cooperation.

An Agenda for Action

Cascadia regional opportunities range from simple information exchange to complex issues management processes. Suggestions for pursuing these opportunities follow.

Sustainable Urbanization

- Regular meetings among regional metropolitan governments to discuss issues such as environmental quality, land use, transportation planning, and governance. An initial meeting, hosted by the Greater Vancouver Regional District, with Portland METRO and the Puget Sound Regional Council, was held on March 31, 1994.
- Regular meetings of government agencies from areas adjacent to the border crossings to discuss border issues affecting their local communities. Under the aegis of the Cascadia Transportation/Trade Task Force and PNWER,

the Whatcom County Council and the City of Surrey have recently formed the Border Crossing Working Group.

- Joint conferences between provincial and state municipal organizations, such as the Union of BC Municipalities and the Association of Washington Cities to explore such common issues as waste management and regional planning.

Regional Advocacy

- Promoting the interests of the Pacific north-south trade corridors, at all levels of government, in competition with other corridors, to attract increased resources and new technology to support the growing cross-border movements of goods and people
- Gaining U.S. federal highway funding for a bi-national pilot project, sponsored by the Cascadia Transportation/Trade Task Force and PNWER, to improve the movement of goods and people along the corridor through infrastructure improvements, operational efficiencies, and bi-national planning and data collection
- Supporting an active marketing program designed to expand participation in PACE on both sides of the border
- Securing broad-based, cross-border community support for renewed Amtrak service between Seattle and Vancouver and uniting as a coalition in support of long-term U.S. federal high-speed rail authorization and funding

Tourism Promotion

- Supporting joint Cascadia regional initiatives that offer potential in the development of tourism products and overseas markets in the following sectors: the Alaska cruise trade, northern partnerships (Alaska, the Yukon, and British Columbia), regional rail touring, Seattle/British Columbia Golden Triangle (Vancouver, Whistler, and Victoria), aboriginal tourism, an ASEAN pilot project, and an intermodal transportation pass (a Tourism Leaders' Forum was held on March 30, 1994, to consider these initiatives)

Trade and Economic Development

- Organizing a regional trade mission on a pilot project basis with private and public sector participation into a third market with growth potential for regionally based industry sectors
- Identifying new opportunities under NAFTA in the Mexican market for joint Cascadia trade-tourism ventures

- Employing the technology of Internet and developing a Cascadia information system with a common database, which would be accessible to regional subscribers through computer bulletin boards

Environmental Cooperation

- Establishing, as a top priority, a bi-national program to repair and preserve the Georgia Basin, the second most threatened ecosystem in the region, through a multiparty process similar to the Fraser Basin Management Program, with 5-year funding committed by all levels of government

Institutional Change

- Mapping the regions' political systems, its institutions, leadership, and organization to produce a regional governance reference guide
- Developing a consultative process among key decision makers in each jurisdiction for the sharing of experience in the creation of new mechanisms for institutional change
- Exploring the potential for expanding and enhancing current bi-national organizations

As Schell and Hamer (1993) point out,

> Cascadia is a recognition of emerging realities, a way to celebrate commonality with diversity, a way to make the whole more than the sum of its ports. Cascadia is not a state, but a state of mind. But a state of mind can have important practical consequences. (Schell & Hamer, 1993, p. 4)

Next Steps

This chapter has offered a snapshot of current regional initiatives and priorities and has raised questions to be answered if regional opportunities are to be realized within a broader, long-term vision. Implementation of an agenda depends on the will of the regions' leaders, their ability to achieve a degree of consensus on priorities, and their willingness to commit resources to addressing them.

Successful implementation also depends on the creation of institutional mechanisms for cross-border collaboration. Some of these initiatives can be undertaken by existing agencies with relative ease, whereas others are more complex and require government agreements, action by bi-national groups, and new or redesigned institutions. This chapter has demonstrated that a wide variety of individuals, groups, organizations, and governments have accomplished a great deal in a relatively short period. It is critical that this energy and drive be sustained and channeled into targeted action. Following are several practical suggestions:

1. A trilateral accord among British Columbia, Washington, and Oregon is required to provide the public policy direction and coordination for implementation of a regional agenda.
2. Federal and provincial/state governments should designate a functional or regional minister/department with primary responsibility for Cascadia regional issues.
3. The Cascadia Transportation/Trade Task Force should establish a bi-national executive committee with a clear mandate to direct the implementation of priority projects.
4. A public policy framework and a set of tools should be produced for use by governments, nongovernmental organizations, and the private sector as a reference base for Cascadia initiatives.

A trilateral accord is an ambitious undertaking but is essential if an institutional framework is to be developed for bi-national action. The existing bilateral relationship between British Columbia and Washington State is not sufficient to address the larger bioregional interests and opportunities of "Main Street" Cascadia.

Political realities must also be recognized. Washington State and Oregon need a joint strategy and alliance when they compete for attention and resources in Washington, D.C. British Columbia has a distinct role to play in cooperating in such initiatives as an integral player in the Cascadia Corridor. Fundamentally, either Oregon and British Columbia cooperate in a trilateral forum, or Washington State has no alternative but to develop a northern and southern strategy for the corridor. The preferred option for all would be shared leadership by the governors of Oregon and Washington and the premier of British Columbia.

NOTES

1. The B.C. Round Table on the Environment and the Economy is an appointed body that reports to the premier of British Columbia on matters relating to sustainability. It issued its report on *The Georgia Basin Initiative: Creating a Sustainable Future* in May 1993. Included with this Round Table report was a background paper (see Artibise & Hill, 1993).

2. One indicator of growth is that the Canada-United States border crossing at Blaine had 11 million crossings in 1990. This was more than double the 1980 figure. Similarly, Vancouver and Seattle-Tacoma were tied in having the fourth highest growth rate (12%) among all metropolitan areas in the United States and Canada between 1985 and 1990. See B.C. Round Table (1993) for detailed statistics.

INTERVIEWS (CANADA)

Earl Anthony	Environment Canada
Lynn Bailey	BC Ministry of Environment, Lands and Parks
Roger Bull	Pacific Northwest Economic Region
Joyce Brookbank	Vancouver Island Tourism Association
Karen Calderbank	BC Trade Corporation
Ken Cameron	Greater Vancouver Regional District
Barry Carlsen	BC Technology Association
Michael Clark	BC Trade Corporation
Lee Doney	BC Round Table on the Environment and the Economy
Tony Dorcey	Fraser Basin Management Program
Gerard Ferry	Capital Region District
Greg Halsey-Brandt	Greater Vancouver Regional District
Peter Heap	Premier Harcourt's Office
George Hunter	BC Biotechnology Association
Bill Jordan	Capital Region District
Ed Kargl	Fraser River Harbour Commission
Erik Karlsen	BC Ministry of Municipal Affairs
Paul Ouimet	Western Transportation Advisory Council
Boris Pavlov	Transport Canada
Norman Stark	Vancouver Port Corporation
Joe Stott	Greater Vancouver Regional District
Richard Taylor	Union of BC Municipalities
Andre Villeneuve	Canada Customs
Nicholas Vincent	BC Ministry of Employment and Investment
Brian Walliser	BC Ministry of Municipal Affairs
Rob Weston	BC Trucking Association

INTERVIEWS (UNITED STATES)

Charlie Allcock	Portland World Trade Center
Rick Bender	Washington Labor Council
Bill Black	Department of Interior, Bureau of Indian Affairs
Mike Brennan	Whatcom Chamber of Commerce
Bill Brubaker	Puget Sound Regional Council
Tom Campbell	Washington Trade and Economic Development
Mark Challender	Whatcom County Council of Governments
Mark Clemmons	Portland Development Commission

Bud Coffey	Boeing Company
Rick Daniels	Washington State Department of Transportation
Don Forbes	Oregon Department of Transportation
Glenn Ford	Oregon Economic Development Department
Gary Grant	Port of Seattle
Fred Hansen	Oregon Department of Environmental Quality
Pat Jones	Washington Public Ports Association
Vera Katz	Mayor of Portland
David Lohman	Port of Portland
Mike Lowry	Governor of Washington
Don Lorentz	Port of Seattle
Gil Mallery	Washington State Department of Transportation
Nancy McKay	Puget Sound Water Quality
Connie Niva	Washington State Transportation Commission
Pat O'Malley	Port of Tacoma
Gary O'Neal	Environmental Protection Agency
Nina O'Neill	Port of Everett
Eileen Quigley	Municipal League of Seattle
Norm Rice	Mayor of Seattle
Henry Richmond	1000 Friends of Oregon
Andrea Riniker	Port of Seattle
Barbara Roberts	Governor of Oregon
Paul Schell	Port of Seattle
Preston Schiller	Sierra Club
Amy Solomon	Northwest Renewable Resource Centre
Bill Stafford	Trade Development Alliance of Seattle and King County
Lucy Steers	1000 Friends of Washington
Jim Street	City of Seattle

REFERENCES

Abbott, C., Howe, D., & Adler, S. (Eds.). (1994). *Planning the Oregon way: A twenty year evaluation.* Corvallis: Oregon State University Press.

Artibise, A. F. J. (1992). *International Georgia Basin-Puget Sound sustainable urbanization project: Rationale and operational plans: Final Report.* Vancouver, BC: Cascadia Planning Group.

Artibise, A. F. J., & Hill, J. (1993). *Governance and sustainability in the Georgia Basin.* Victoria: British Columbia Round Table on the Environment and the Economy.

British Columbia Round Table on the Environment and the Economy. (1993). *Georgia Basin Initiative: Creating a sustainable future* [with Addendum]. Victoria: Author.

British Columbia/Washington Environmental Cooperation Council. (1993, July). *1992-93 Annual Report.* Victoria, BC, and Olympia, WA: Author.

Cascadia Institute. (1994). *Opportunities for expanding international tourism in Cascadia: A Canadian perspective.* Vancouver, BC: Author.

De Grove, J. M. (1992). *Planning and growth management in the states: The new frontier for land policy.* Cambridge, MA: Lincoln Institute of Land Policy.

Fraser Basin Management Board. (1992). *Strategic plan for Fraser Basin management program, 1993-98.* Vancouver, BC: Author.

Greater Vancouver Regional District. (1993a, August). *Livable region strategy: Proposals: A strategy for environmental protection and growth management.* Burnaby, BC: Author.

Greater Vancouver Regional District. (1993b, September). *Transport 2021 report: A long-range transportation plan for Greater Vancouver.* Burnaby, BC: Author.

Greater Vancouver Regional District. (1993c, October). *Transport 2021 report: A medium-range transportation plan for Greater Vancouver.* Burnaby, BC: Author.

Hamer, J., & Chapman, B. (1993). *International Seattle: Creating a globally competitive community.* Seattle, WA: Discovery Institute.

High Speed Rail Development Act of 1994, Pub. L. No. 103-440, 49 U.S.C. § 20101, 108 Stat. 4615.

Intermodal Surface Transportation Efficiency Act of 1991, Pub. L. No. 102-240, 498 U.S.C. § 101, 105 Stat. 1915.

Kelly, C. (1994, Spring). Midwifing the new regional order [Publisher's note]. *The New Pacific*(10), 6.

Ketcham, P., & Siegel, S. (1991). *Managing growth to promote affordable housing: Revisiting Oregon's goal 10.* Portland: 1000 Friends of Oregon and Home Builders Association of Metropolitan Portland.

Northwest Resources for Regional Cooperation. (1990, July). Seattle, WA: Northwest Policy Center.

Oregon State. (1991). *Benchmarks.* Salem, OR: Governor's Office.

Premier's Office (Province of British Columbia). (1993, June 21). *Release of Georgia Basin Report* [Press release]. Victoria, BC: Author.

Roseland, M. (1992). *Toward sustainable communities: A resource book for municipal and local governments.* Ottawa, ON: National Round Table on the Environment and the Economy.

Schell, P., & Hamer, J. (1993). *What is the future of Cascadia?* Seattle, WA: Discovery Institute.

The Lower Fraser River Basin: A state of the environment synopsis. (1992). Vancouver, BC: Environment Canada.

U.S. Department of Transportation, Federal Highway Administration. (1994, January). *Intermodal Surface Transportation Efficiency Act: Section 1089 and Section 6015: Assessment of border crossings and transportation corridors for North American trade: Report to Congress* [News release]. Washington, DC: Author.

Washington State Growth Management Act of 1990, 1991 SP. S. c. 32 § 40.

Washington State. (1989). *Environment 2010: Action agenda.* Olympia, WA: Governor's Office.

11

Developing Global Cities in the Pacific Northwest: The Cases of Vancouver and Seattle

THEODORE H. COHN
PATRICK J. SMITH

Vancouver and Seattle are global cities. Their international involvement is long-standing, and their linkages with foreign cities have expanded dramatically in recent years. Since the 1940s and 1950s, Vancouver and Seattle have also broadened their international objectives to include a wide range of activities in the cultural, educational, economic, environmental, and foreign aid areas. The global cities phenomenon in general has become more important as the scope of international relations has increased, encompassing "new" policy areas such as environmental pollution, human rights, immigration, monetary and trade instabilities, and sustainable development. Unlike traditional strategic-security matters, these new issues are intermestic in nature, that is, they are "simultaneously, profoundly and inseparably both domestic and international" (Manning, 1977, p. 309). Constituent diplomacy of subnational actors at the provincial and municipal levels has confirmed the importance of intermestic politics (see Kincaid, 1990).

This chapter examines the global, national, subnational, and idiosyncratic (or individual) determinants of municipal international policy making. In assessing the relative importance of these determinants, an examination of paradiplomatic policy phases in the Pacific Northwest cities of Vancouver and Seattle forms the primary basis for analysis. These two cities are key centers in the Georgia Basin-Puget Sound region, which is emerging as a major bi-national global region.[1] Before discussing determinants, we first present a typology that illustrates the intermestic interactions of municipal governments in the United States and Canada. The typology demonstrates that determinants are not unidirectional and also that municipal governments can be mediating as well as primary actors.

A Typology of Municipal Internationalism

The growth of interdependence highlighted the shortcomings of the realists' state-centric vision of global politics, and new concepts were introduced to account for a more diverse group of actors. Keohane and Nye (1976) have defined *transgovernmental relations* as "direct interactions between agencies (governmental subunits) of different governments where those agencies act relatively autonomously from central governmental control" (p. 4). Although Keohane and Nye viewed transgovernmental relations as occurring primarily among bureaucratic subunits of national governments, their definition may also include provinces, states, or cities that act relatively autonomously from senior governmental control. Cities are "hybrid international actors" because they lack the qualities of sovereignty normally associated with higher levels of government. Nevertheless, cities engage in transgovernmental relations because they are governmental subunits that "are linked by various means into the network of governmental and administrative processes" (Hocking, 1986, p. 483). Transnational relations are "interactions across the border in which at least one actor is nongovernmental" (Keohane & Nye, 1976, p. 4). This definition can be used to describe cross-border interactions of a municipal government with nongovernmental actors such as transnational corporations.

Our "Global Cities Typology 1" (see Figure 11.1) illustrates the transgovernmental and transnational interactions of municipal government "A" as a *primary actor.* Subnational governments are primary actors when they engage *directly* in global relations. As Typology 1 shows, a city's cross-border interactions may be *egressive flows,* which are initiated from the inside out or *ingressive flows,* which are channeled from the outside in.[2] Alternatively, the flows may occur in both directions simultaneously. A city government (city "A" in typology) may have cross-border relations with governmental and private actors at the international, national, provincial/state, and municipal levels. Domestic determinants of a city government's global interactions stem from horizontal linkages with other city governments (city "B") and from vertical linkages with more senior governments at the federal and provincial/state levels (as well as from a changing public policy agenda).

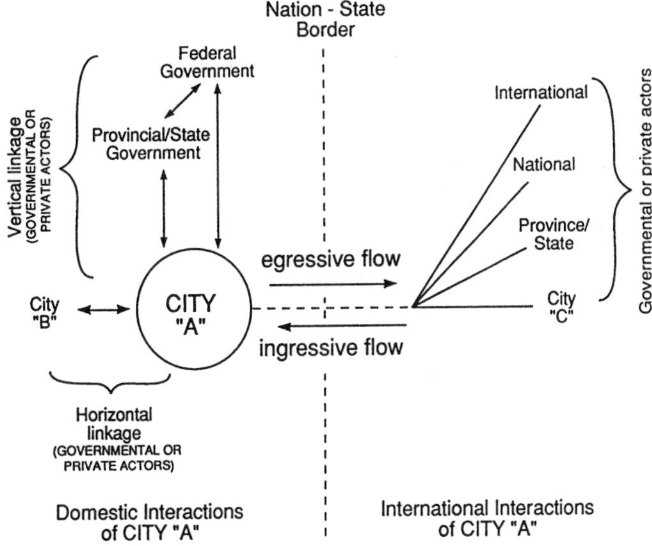

Figure 11.1. Global Cities Typology 1

SOURCE: From Smith and Cohn, "International Cities and Municipal Paradiplomacy: A Typology for Assessing the Changing Vancouver Metropolis" (p. 621), 1994, in F. Frisken (Ed.), *The Changing Canadian Metropolis: A Public Policy Perspective* (Vol. 2), Berkeley: University of California, Institute of Governmental Studies; Toronto, ON: Canadian Urban Institute. Used with permission of the author.
*CITY A represents municipal government.

Our "Global Cities Typology 2" (see Figure 11.2) illustrates the intermestic interactions of municipal government "A" as a *mediating actor.* Mediating actors seek to influence their more senior governments (both provincial/state and federal) for the purpose of promoting "general policies that are beneficial to local conditions in such areas as trade and foreign investment" (Hocking, 1986, p. 484). Typology 2 shows that cities may also interact with each other through horizontal linkages (cities "A" and "B") in efforts to influence provincial/state and federal government foreign policies. Whether a city is a primary or a mediating actor, its domestic and international interactions may be cooperative, competitive, or conflictual in nature.[3]

Municipal governments are more inclined to engage in mediating activity when the policies they desire are clearly out of their jurisdiction or beyond their capacities (e.g., funding requirements) to implement through

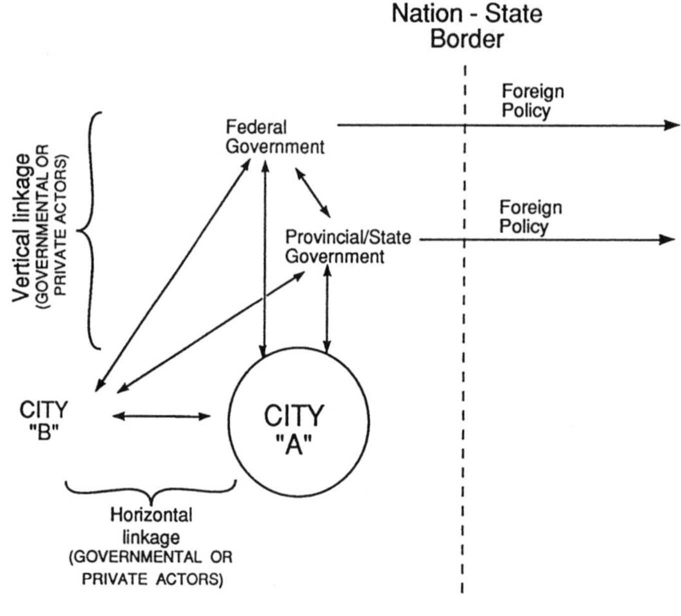

Figure 11.2. Global Cities Typology 2

SOURCE: From Smith and Cohn, "International Cities and Municipal Paradiplomacy: A Typology for Assessing the Changing Vancouver Metropolis" (p. 626), 1994, in F. Frisken (Ed.), *The Changing Canadian Metropolis: A Public Policy Perspective* (Vol. 2), Berkeley: University of California, Institute of Governmental Studies; Toronto, ON: Canadian Urban Institute. Used with permission of the author.
*CITY A represents municipal government.

direct action. Municipal governments are also more willing to choose mediating activity when their basic interests coordinate with those of more senior governments and when their relationship with more senior levels is generally trusting and cooperative. On the other hand, cities often become primary actors when they are dissatisfied with the policies of their senior governments and are unsuccessful in altering those policies through mediating activity. Primary activity is also appropriate for a variety of regional issues that are of more immediate relevance to cities than to more senior levels of government. Primary and mediating activities are, of course, not mutually exclusive, and municipal governments often engage in both types of behavior simultaneously.

We have used the framework provided in Typologies 1 and 2 to conduct a detailed analysis of the intermestic interactions of Canadian and American cities, provinces, and states.[4] Reference to the typology is useful here in examining the determinants of municipal international relations in Vancouver and Seattle.

Municipal Policy Determinants: Preliminary Observations

The determinants of municipal international relations can be categorized as global, national, subnational, and idiosyncratic, although these four levels overlap considerably. A major *international determinant* relates to advances in technology and the associated growth in interdependence. Technological advances have in fact contributed to "an interdependence of local, national, and international communities that is far greater than any previously experienced" (Rosenau, 1990, p. 17). As a result, leaders of subnational governments have become "acutely aware of the influence which international actors . . . can have on the economic well-being of their constituencies" (Fry, 1986, p. 301). The success of these leaders in attracting investment, trade, and tourism can contribute to the economic vitality of their cities and to their reelection. This is increasingly the case in the global policy stances of the cities of Vancouver and Seattle.

Subnational governments have become involved internationally for defensive as well as proactive reasons. Provincial/state and municipal activities that were previously viewed as domestic are now often considered to be "trade-distorting" and thus subject to international scrutiny. For example, the General Agreement on Tariffs and Trade (GATT; 1986) states that "each contracting party shall take such reasonable measures as may be available to it to ensure observance of the provisions of this Agreement by the regional and local governments and authorities within its territory" (Article XXIV.12). The Canada-U.S. Free Trade Agreement and the North American Free Trade Agreement (NAFTA; 1992) contain even stronger statements requiring the member states to ensure observance by their constituent units. Thus, NAFTA stipulates that "the Parties shall ensure that all necessary measures are taken in order to give effect to the provisions of this Agreement, including their observance . . . by state and provincial governments" (Article 105). This statement is also applicable to municipal governments because NAFTA specifies that "for the purposes of this Agreement . . . a reference to a state or province includes local governments of that state or province" (Article 201.2).

Not surprisingly, many state, provincial, and municipal governments in North America have expressed concerns that NAFTA will delimit their freedom to establish policies dealing with such issues as the environment, government procurement, and subsidies. In Canada, the provincial governments have substantial responsibility for many labor and environmental matters. It is likely that several provinces (most notably, the New Democratic Party-led provinces of Ontario, British Columbia, and Saskatchewan) will be unwilling to subject their laws to the trilateral process established in NAFTA's environmental and labor side accords (see Fagan, 1993). In the United States, the Center for Policy Alternatives (1993) has compiled a detailed listing of resolutions by state and municipal governments and more specialized interests regarding NAFTA. For example, the city of Milwaukee, Wisconsin, passed a resolution expressing "the Common Council's strong opposition to NAFTA because of its potentially adverse impact on employment, economy and property tax base" (p. 8). Cities sometimes have adopted common cause with their state governments, as Los Angeles did when it passed a resolution indicating (in part) that "NAFTA must not preempt state authority" and that "the agreement should also utilize public input in its formation" (p. 8). State-based groups in turn have sometimes been supportive of local as well as state governments. Thus, the Democratic Party of the state of Washington passed a resolution warning that "NAFTA could preempt state and local laws" (p. 9). It should be noted that some reactions by municipal governments to NAFTA have been positive (often with qualifications). For example, the U.S. Conference of Mayors passed a resolution endorsing NAFTA, but it also requested that strong supplemental agreements be established.

Geographic location is another international determinant of municipal international relations. Geographic factors may be less important when a municipal government engages in global paradiplomacy, which deals "with issues concerning the whole international system (e.g., peace and war and liberalization of international trade)" (Soldatos, 1990, p. 37). Nevertheless, it is more common for cities to engage in regional paradiplomacy, which concerns issues "of a regional relevance to the communities taking part in a subnational activity" (p. 37). Soldatos identifies two types of regional paradiplomacy: macroregional, which involves noncontiguous communities (e.g., Vancouver and Tokyo), and microregional, which involves a degree of geographical contiguity (e.g., Vancouver and Seattle).

As we discuss later in this chapter, the location of Vancouver and Seattle as port cities in the Pacific Northwest has a major influence on their international activities. For example, both cities have increased their linkages with a number of Asian cities and states on the Pacific Rim, and microregional ties between Seattle and Vancouver are also solidifying in areas such as trade, transportation, and the environment.

National determinants of a city's international relations include the declining capability of central governments to provide solutions for many current policy problems and the generally supportive attitude of national governments toward municipal internationalism. The emergence of cities as significant global actors contradicts many of the traditional views about the natural order of governance, whereby, constitutionally, foreign policy making has been a primary responsibility of national governments. Nevertheless, national governments do not normally disapprove of municipal involvement in international matters, and they often even encourage and facilitate such involvement.

Federal officials, of course, frown on certain city activities that seem to contradict the objectives of the national government; these have included Seattle's decision to establish sister city linkages with Managua (Nicaragua) and the moves by Winnipeg (Manitoba) and Maple Ridge (British Columbia) to link with cities in Taiwan. Cities generally lack jurisdiction, however, in international matters, and therefore they do not interfere significantly with the foreign policy activities of central governments. Although a municipal government's establishment of a particular sister city linkage or of a nuclear weapons free zone may create some resentment, such actions are usually considered to be largely symbolic and do not pose a major threat to national priorities. Thus, the Office of International Affairs in Seattle maintains that "the role of local government at the international level is clearly distinguished from the role of the federal government."[5] The global activities that most preoccupy local officials are in the areas of promoting trade and foreign investment, and the federal government tends to be fully supportive of these activities. Canada's External Affairs Department, for example, has assisted Vancouver with trip arrangements to its sister city of Odessa, and the U.S. State Department has provided Seattle with foreign service officers for 1- to 2-year periods through the Pearson fellowship program.

Another important national determinant of municipal internationalism relates to the declining capability of central governments to find adequate solutions for many major policy problems. This incapacity stems partly from the growing importance of new intermestic issues that are not completely within a nation-state's jurisdiction (e.g., environmental pollution, terrorism, currency crises, and AIDS) and partly because older domestic issues are increasingly linked with international components. Examples of such older issues include agriculture, textiles, and labor, which have contributed to major controversy over trade liberalization efforts at both the GATT and NAFTA negotiations (e.g., Cohn, 1993). Because citizens are dissatisfied with the problem-solving capabilities of their central governments, they often look to subnational governments at the provincial/state and local levels for solutions.

The possession of sovereignty is normally viewed as a strength of the nation-state, but sovereignty also poses some major limitations. Indeed, Rosenau (1990) refers to nation-states as sovereignty-bound actors and to transnational corporations and subnational governments as sovereignty-free actors. This terminology "serves as a continual reminder that the differences between states and other collectivities may not be as one-sided as they are usually assumed to be" (p. 36). In the 1940s, the compromise of embedded liberalism differed from orthodox liberalism in that it prescribed "state action to contain domestic social and economic dislocations generated by markets" (Krasner, 1983, p. 4; see also Ruggie, 1983). This change added to the burdens on central governments, and their responsibilities continued to increase as the number of global issues and actors multiplied. Today, a nation-state's sovereign responsibilities can overwhelm its capacities and prevent the state from concentrating its energies on specific policy objectives. Sovereignty-free collectivities such as subnational governments, by contrast, need not disperse their responsibilities as widely and can direct more resources to a select group of preferred goals. The shifting pattern demonstrated in the policy phases of Vancouver's and Seattle's international activities (discussed later in this chapter) highlights this point.

The declining problem-solving capability of nation-states (although not of their formal constitutional authority) has induced subnational governments to adopt increased responsibilities and functions.[6] Local communities and individuals who feel the effects of interdependence have often turned to their municipal and provincial/state governments when more remote central governments have failed to meet their needs and interests. This tendency can be referred to as subgroupism, which contrasts with the earlier centralizing tendencies of nation-statism. Some analysts, such as Seelig and Artibise (1991), write that subgroupism and interdependence are important in explaining the growth of municipal internationalism: "Traditional boundaries between countries . . . [have] become increasingly meaningless . . . cities [have] become independent of their countries and deal directly with other parts of the world, [and] city and regional interests . . . will be far more important than national and provincial economic policies" (p. 75).

Although the role of subnational units has increased partly as a result of disillusionment with central governments, it is important to note that citizens' expectations can also undermine the authority of provinces, states, and cities. The growth of interdependence is seemingly inexorable, and subnational as well as central governments are having difficulties in dealing with the impact of global actors and events on local communities. Thus, progressive movement toward transnationalism could be accompanied by a further fracturing of subgroups (Rosenau, 1990). The marked proliferation of ethnic conflict today is attributable in part to these dual tendencies toward transnationalism and subgroupism.

Despite the problems confronting national governments, they continue to have legitimacy as international actors and substantial material resources, which can be useful to municipal governments in promoting their own international activities. City governments may therefore vie for the attention of the federal government in efforts to raise their international profiles. An example was the decision of the new (1984) Progressive Conservative government in Ottawa to establish Vancouver and Montreal as international banking centers. In efforts to maintain its preeminence as a Canadian financial center, Toronto pressured the federal government to reverse its policy in this area. Toronto could not prevent the government from passing the necessary legislation, but it did succeed in weakening the provisions and thus in limiting Vancouver's and Montreal's competitive advantages (see Goldrick, forthcoming).

Although municipalities often engage in mediating activity, they may shift to primary activity and become more directly involved internationally when they feel that the federal government is not sufficiently attuned to their interests. This is often the case for cities further removed from their national capitals, such as Vancouver and Seattle. Thus, a former premier of Alberta maintained that "many Western Canadians are tired of going to Asian countries and meeting with well-meaning but Eastern-Canadian-oriented civil servants who can relate the name of every major company doing business in Montreal or Ottawa or Toronto, but who have never heard of some of the international concerns of Winnipeg, Regina, Edmonton, Calgary or Vancouver" (quoted in Johannson, 1978, p. 364).

Subnational determinants of municipal international relations may be constitutional, political, economic, and cultural. An important subnational determinant stems from the relationship between cities and their more senior state or provincial governments. The emergence of cities as significant global actors runs counter to the derivation of local authority in the United States and Canada from a senior level of government: "At common law, and under the constitution, Canadian and American local governments . . . are . . . 'tenants at will' of the provinces or states" (Jones, 1988, pp. 89-90). Although there is considerable opportunity for compromise on intergovernmental policy matters in the American institutional setting, senior constitutional authorities in Canada are more inclined to view any significant policy divergence as conflictual. Such perceived policy conflicts naturally have their jurisdictional dangers. Instances of a senior government exercising its "nuclear option" serve as reminders to local authorities about "when, how and where . . . [senior] governments exercise their will against their recalcitrant or innovative children" (p. 91; see also Smith, 1986, 1988).

Despite the possibilities for retribution, it is intriguing that municipal international policy innovations have engendered little conflict to date. As is the case with municipal-federal government relations, the current Vancouver mayor, Gordon Campbell (interview, 1988), has stated that the city "is not curtailed in any way at all by the province." Senior city staff have pointed out that the province in fact often encourages and facilitates Vancouver's international involvement: "When we were in Tokyo last year, we went and met with the Provincial Government House in Tokyo, and they were very helpful in terms of providing us with information and assistance" (interview, T. Droetboom, 1988). Some city officials maintain that Vancouver's special position gives it an added opportunity for international activity: "Certainly, the City of Vancouver, although it is a creature of the province, has never felt that to the same extent that any other city in B.C. does" (interview, M. Kinsella, 1988).

Another subnational determinant relates to the learning process of municipal governments as a result of their intergovernmental and global activities. The skills and experience that cities gain from intergovernmental relations with their senior governments serve them well internationally, and city involvement in global activities in turn instills the desire and confidence to engage in further forays into the international realm. Nevertheless, international involvement is still not considered to be an ordinary activity for most subnational units, particularly for municipal governments. Although cities have gained considerable international experience and contacts through the years, they tend to lack funding for global activities, and the available funding is often an early victim of municipal budgetary cutbacks. For example, when fiscal restraint returned in the 1980s, Vancouver slashed the budget for its strategic cities program by two thirds, and Seattle severely cut back the number of staff involved with international activities (see following discussion).

Subnational determinants may also be sociocultural in nature. The growth of ethnic diversity in major North American cities provides both the incentive and opportunity for establishing international linkages. This ethnic diversity is evident in Seattle and to an even greater extent in Vancouver. The 1990 census data show that 11.2% of Seattle residents were Asian, with the largest Asian groups being Chinese, Filipino, Japanese, Vietnamese, and Korean. The Asian population in Vancouver is also highly significant in both numbers and influence. In 1992, there were 29,013 foreign immigrant landings in Metropolitan Vancouver (and 35,459 in British Columbia). Asia was by far the most important source of immigration, accounting for 77.5% of the landings. Hong Kong alone accounted for 31.3%. With regard to foreign languages, the 1991 census reports that only 59.8% of Vancouver residents identified English as their first language, whereas 18.5% listed their first language as Chinese (City of Seattle, Division of International Affairs, 1992; City of Vancouver, Economic Development Office, 1993).

Idiosyncratic determinants are the most difficult to generalize about, but they can, nevertheless, have a significant impact on the international behavior of subnational governments. The personal leanings and preferences of Vancouver and Seattle mayors sometimes have considerable influence on both the level and nature of municipal international involvement. A major reason for the importance of idiosyncratic determinants relates to the sovereignty-free nature of subnational units. Without the full complement of sovereign responsibilities borne by the central government, municipal leaders can focus their energies (and their idiosyncrasies) on international issues of special interest to them. Some of these personalized issues are discussed below in the next section of the chapter.

Policy Phases and Policy Determinants

This portion of the chapter traces the determinants of international policy making (and policy-making capacity) in Vancouver and Seattle through an analysis of four phases. The first phase was ad hoc in nature and extended from the 1940s to the early 1970s. The ad hoc phase was indicative of relatively immature policy intent and capacity and was oriented primarily to cultural and educational exchanges. The second phase, marked by efforts to develop a more rational approach to international activities, extended from the latter part of the 1970s to the mid-1980s. The rational phase emphasized the development of an institutional capacity and a greater focus on economic linkages.

The third phase focused on the development of a more strategic internationalist policy position in the late 1980s. Municipal paradiplomacy in the strategic phase began to place much greater emphasis on a business-oriented economic development approach. Thus, Vancouver shifted from adding more formal sister city twinnings (still limited to five) to establishing less formal linkages with cities that were identified as gateways to significant regional and national economies. In Seattle, the strategic phase involved a significant expansion of formal sister city twinnings, which now number 20. The fourth phase, which we will term a globalist policy stance, is more comprehensive than the earlier phases and is still in the process of developing. The globalist phase represents a significant expansion in policy content from a cultural-educational and business-economic focus to additional emphases on world peace and disarmament, international aid, and environmental-sustainability components. More than in previous phases, the globalist orientation indicates that municipal governments are no longer prepared to passively wait for senior governments to take action on pressing global matters. The municipalities instead are becoming directly involved in a variety of issues to counter threats to humanity (e.g., environmental degradation) and to improve the quality of life for individuals.

Although we discuss the international policy making of Vancouver and Seattle using these four distinct phases, in reality the phases are overlapping. Each successive phase also incorporates the characteristics of the previous phases. For example, although the strategic phase emphasized business and economic benefits, cultural exchange (from the ad hoc phase) was seen as contributing significantly to economic development opportunities. In the globalist phase, economic exchange continues to be a central aspect of municipal international activity. The municipality, however, also adopts a much broader policy perspective, reflecting more clearly the stance of a municipal global citizen (see Alger, 1987). The gradual change from incremental to more rational approaches in the four phases is indicative of a growing institutionalization and maturing of subnational policy making in the global arena. Each policy phase has reflected a different set of choices by Vancouver and Seattle in keeping with shifting policy determinants.

The Settings

Seattle[7]

Seattle is the largest city in Washington State, with a population of 516,259 in 1990. It is located in King County (pop. 1,507,300) and forms the central core of the Puget Sound Basin, where nearly two thirds of the state's population (of almost 4.9 million) resides. Some of the largest employers in the Seattle-Puget Sound area are important contributors to the international focus of the region. The Boeing Company, for example, is the largest aircraft manufacturer in the world, providing over half of the manufacturing jobs in the Seattle area. About two thirds of Boeing's sales (totaling over $17 billion in 1988) are directed to foreign countries. Microsoft, the largest personal computer software company in the world, is the second largest employer among public corporations in the Seattle area and also has a substantial volume of foreign sales. The U.S. Defense Department is yet another major employer in the Puget Sound region with important international linkages and responsibilities.

Trade statistics are another indication of the degree to which global interdependence affects Seattle and Washington State. In 1990, exports originating in Washington accounted for 23.5% of gross state product, and the state's exports per capita were valued at $5,058. This per capita figure was higher than for any other U.S. state and twice the U.S. average. The top 15 markets for Seattle-King County exports in 1990 were all European and Asian except for Australia (ranked 4th), Canada (ranked 5th), and Brazil (ranked 14th). Furthermore, 6 of the top 15 markets for Seattle-King County exports in 1990 were Asian, reflecting the city region's strong Pacific focus. Seattle's dependence on trade is also evident from its role as a major port city. The total value of imports and exports through the Port of Seattle in 1990 was $28 billion, and Seattle serves as the top U.S. port in container tonnage exports to Asia (3.9 million metric tons in 1990). Seattle also serves as an important conduit to and from Asia for "people traffic" because it is the closest gateway to Asia in the continental United States (9 hours by air from Tokyo).

Tourism, foreign investment, and foreign-owned firms and banks are other important indicators of Seattle's international activity. Tourism is Washington State's fourth largest industry, generating $5 billion in revenue in 1989, much of it centered in Seattle (e.g., Washington State's Convention and Trade Center opened in downtown Seattle in 1988). The top six countries providing foreign tourists to Washington (in order of importance) are Canada, Japan, Britain, Taiwan, Hong Kong, and Germany. Foreign investment in Washington State and Seattle is difficult to estimate because foreign investors are not required to register with any agency other than the Internal Revenue Service, and this information is kept confidential. On the basis of the amount of employment generated, the most significant investors in Washington State are Canada, Japan, Germany, Switzerland, New Zealand, Britain, and Hong Kong. A recent study of Greater Seattle has identified five countries (four of which are Asian) as providing the best opportunities for attracting further foreign investment to the region: South Korea, Taiwan, Singapore, Hong Kong, and Canada (Trade Development Alliance of Greater Seattle, 1992). About 200 foreign-owned firms employ more than 10,000 people (approximately 3% of the workforce) in Seattle, and 14 banks headquartered abroad have Seattle offices.

The Trade Development Alliance of Greater Seattle (1992) has conducted a detailed analysis of the international business relationships of the Seattle-King County region. The alliance is a partnership of the City of Seattle, the Port of Seattle, King County, organized labor, and the Greater Seattle Chamber of Commerce. In identifying "the top current customers" for the Greater Seattle-King County region, the alliance rated countries according to six criteria: exports without transportation, transportation exports, marine port, air cargo, investment, and tourism. A composite ranking on the basis of these six criteria indicates that the following countries are the city-region's top current customers (in order of importance): Japan, Canada, Britain, China-Hong Kong, Germany, Australia, South Korea, and Taiwan. For the most part, these countries are located in Asia and to a lesser extent in Europe. The study authors, however, clearly view Canada as "unique" because it is a part of the Pacific Northwest microregion:

> Among our top customers, Canada, number two, is unique. In many ways it has the characteristics of a domestic market. While we recognize the importance of Canada as a foreign trading partner of this region, we do not want to ignore the unique features of Greater Seattle/King County's economic links to Canada. (Trade Development Alliance of Greater Seattle, 1992, p. 6)

Vancouver

The Vancouver metropolitan area is the third largest in Canada with a population of 1,643,451 in 1992; the population of the City of Vancouver (as of 1992) is 477,872. Vancouver is located on the Pacific Coast in British Columbia, Canada's third most populous province (pop. 3,312,400). Despite the large size of British Columbia (slightly larger than the states of Washington, Oregon, and California combined), about 55% of the population resides in the "Lower Mainland," composed of four regional districts along the Fraser River adjacent to Vancouver. Politically, the region elects slightly more than half of the 75 members of the provincial Legislative Assembly (see Oberlander & Smith, 1993).

Trade statistics clearly demonstrate the degree to which British Columbia and Vancouver are affected by global interdependence.[8] International exports are directly responsible for more than one in five British Columbia jobs and for about one quarter of the province's gross domestic product. Although exports of value-added products have increased in recent years, they accounted for only 11% of British Columbia's total exports in 1991; thus, the province is still principally an exporter of semiprocessed and unprocessed natural resource products. Forest products continued to constitute more than half of the total value of British Columbia's exports in 1992, with minerals and energy products accounting for another 18%. In the decade of the 1980s, British Columbia's exports grew at a real average annual rate of 4.6% and were valued at $17.8 billion in 1989. Provincial exports, however, fell by 9.2% in 1990 and by 8.6% in 1991. The falling export earnings, which reflected declining international commodity prices and a slowdown in economic activity in major markets, were a stark reminder of British Columbia's vulnerability to conditions in the global economy. Exports began to increase again in 1992, largely because of a surge in sales to the United States.

Since the end of World War II, the country profile of British Columbia's exports has changed dramatically. Although dependence on the American and British markets has declined, the share of exports to Asia and the European Union (EU) has generally increased. The province of British Columbia exports proportionately far more to the Asia Pacific and the EU and less to the United States than the Canadian average. In 1990, about 42% of British Columbia's exports were directed to the United States, 38% to the Pacific Rim, and 15% to the EU. The figures for total Canadian exports by contrast were 74% to the United States, 11% to the Pacific Rim, and 8% to the EU. It is important to note, however, that the province's diversified trading pattern is limited mainly to resource commodities. The United States was by far the major destination for British Columbia's *value-added* exports in 1990, taking about 70% of the total.

A large proportion of British Columbia's exports to the United States are also concentrated regionally. Indeed, more than half of the province's total exports to the United States is destined for five Western states:

Washington, Oregon, California, Montana, and Idaho. In 1991, the three Pacific Northwest states (Washington, Oregon, and Idaho) combined purchased more British Columbian exports than all of Europe, and Washington State alone imported over three times the amount of British Columbian goods imported by either Britain or Germany. Thus, a high degree of trade interdependence exists within the Pacific Northwest microregion. Within the Pacific Rim, Japan is British Columbia's most important trading partner, taking about 76% of the province's total exports to that region; the Republic of Korea replaced Britain as British Columbia's third most important single export market in 1991. The movement of goods through the Port of Vancouver is another indication of the city's strong Pacific focus. Thus, the nine top export destinations and six top sources of imports through the Port of Vancouver in 1992 were all Asia-Pacific countries.

British Columbia's dependence on foreign investment has also increased in recent years. From 1985 to 1991, British Columbia accounted for 12% of all foreign investment transactions in Canada. The major sectors of foreign investment activity in British Columbia in value include manufacturing (39%), wholesale and retail trade (22%), real estate (15%), and tourism services (9%). As with trade, British Columbia is relatively more dependent on Pacific Rim investment and less dependent on American investment than the rest of Canada. In 1988, British Columbia accounted for 37.7% of revenues earned by Pacific Rim-controlled corporations in Canada, far in excess of the relative size of the British Columbia economy (Duncan, 1992; Statistics Canada, 1988; see also Thorne, 1993).

The background data on Seattle and Vancouver demonstrate that both cities have developed particularly close economic and cultural linkages with Asia-Pacific countries and with the Pacific Northwest microregion. The data also show that both cities have become highly dependent on foreign investment and trade. These characteristics will help to explain the motivations for international activity of the two cities in the policy cases that follow.

Vancouver Policy Phases

The Ad Hoc Policy Phase: The 1940s to the 1970s

Vancouver was the first North American city to establish a sister city relationship when it twinned with the (then Soviet-allied) city of Odessa in 1944. The local rationale was rooted in humanitarian assistance to a war-devastated sister port. Although Cold War relations after 1945 limited further contact, this sister city link was never broken. To this point, the linkage has been largely cultural, for example, in facilitating ties between the Jewish communities in the two cities (Smith, 1992). Vancouver added the sister cities of Yokohama in 1965 and Edinburgh in 1978. The incremental pattern of establishing relationships in each case was similar to that for Odessa: intense initial involvement by parts of the community leading to the establishment of formal municipal linkages, followed by lengthy periods of relative neglect. Vancouver alderwoman Libby Davies (interview, October 6, 1988) has aptly described the sporadic nature of the city's global exchanges in the incremental phase, stating that there was "never a set program; it was just a matter of evolution." Depending on the interest of a particular mayor, city councilor, or segment of the community, there would be brief bursts of interaction followed by more extensive periods of inactivity.

Besides being sporadic in nature, these early international exchanges were based primarily on cultural and educational linkages. Vancouver's incremental policy phase was fully consistent with the objectives of the Federation of Canadian Municipalities (FCM; 1988) for sister city twinnings: to provide direct contact between diverse peoples and thus foster international understanding; to expand contact between homelands of new Canadians and Canadian communities; to develop an appreciation of foreign culture, history, and traditions; and to develop better perspectives on problems/opportunities at home. The cultural and ad hoc bases for Vancouver's global interactions in the incremental phase were entirely natural for some city officials and community leaders, but by the end of the 1970s, others viewed these relationships as clearly "not enough" (interview, Sid Fancy, economic development manager, City of Vancouver, October 27, 1989).

The Rational Policy Phase: 1980 to 1986

In the rational policy phase, Vancouver's international activities began to shift toward more economically oriented objectives in response to the growth of interdependence in trade, foreign investment, and other areas. The election of Mike Harcourt in 1980 (with substantial leftist support) was also an element in the policy shift. During his three terms as mayor (1980 to 1986), Harcourt sought to establish the city's internationalist links on a less ad hoc basis. Besides participating actively in the FCM and in the discussions of "Big City" mayors, Harcourt also promoted the case of

Vancouver as Canada's Pacific Gateway. In accordance with the new rational policy emphasis, the city's Asia-Pacific focus was based on economic as well as cultural considerations.

The Harcourt government engaged in a combination of primary and mediating activity to promote its international policies. It worked closely with the new (1984) Conservative federal government to establish Vancouver as one of two Canadian international banking centers and to strengthen Vancouver's Pacific Rim links. It also revitalized the Yokohama sister city linkage in efforts to upgrade economic ties with Japan. In 1985, Vancouver added Guangzhou, China, as a sister city. Although the Harcourt government relied on cultural linkages with the large Vancouver Cantonese community in establishing the Guangzhou linkage, the importance of Guangzhou's economic ties with Hong Kong was a primary consideration. Vancouver's choice of Los Angeles as a U.S. twin in 1986 was yet another indication of the city's new, more rational focus. Despite proposals for many other U.S. relationships, the Vancouver Board of Trade joined with Harcourt in pushing for Los Angeles because of business opportunities in the sprawling metropolis, particularly in the film industry. Vancouver's economic development manager Sid Fancy (interview, October 27, 1989) described the objectives of the rational policy phase, stating that a "sister city program has to focus on something" and that the Harcourt government was concentrating "on economics" with a "Pacific focus."

The more rational policy stance was also reflected in the organization and funding of Vancouver's international initiatives between 1980 and 1986. Before the Harcourt period, there was no organizational center for Vancouver's global activities, and the city allocated funds for these activities on an ad hoc basis. To provide more consistency, Mayor Harcourt asked the city council to establish an annual budget for Vancouver's international program in 1982. Initially set at $30,000, this budget grew to $100,000 by 1985 (out of a total city budget of $300 million). Despite the limited nature of this budget, it served as seed funding that had substantially greater impact because of private sector multipliers (e.g., hosting dinners and arranging air travel) and senior governmental grants (e.g., travel funding for cultural and artistic groups to participate abroad). Organizationally, day-to-day staff responsibility for Vancouver's global activities shifted to the city's economic development office, whereas protocol responsibilities were placed in the city clerk's office; the mayor's office also became directly involved when needs dictated.

The success of Harcourt's more rational city paradiplomacy highlighted the limitations of the incremental phase with its emphasis on formal sister city links. The confluence of increased interdependence, a proactive internationalist mayor, and senior governments with coincident Pacific policy goals created important global opportunities for Vancouver. Furthermore, the Harcourt initiatives in the rational policy phase contained the seeds of the third and fourth phases of Vancouver's international policy making. The third (strategic) phase reflected a more obvious business focus; the fourth (globalist) phase was more comprehensive in nature, emphasizing city contributions on a wide range of issues including foreign aid, world peace, and the environment. Under Harcourt, for example, Vancouver designated itself a nuclear weapons free zone, reflecting a view that peace and disarmament issues were clearly within the city's policy purview. Harcourt was also instrumental in federal foreign aid initiatives such as the programs funded by the Canadian International Development Agency and administered by the FCM. These included the China Open Cities program, the Municipal Professional Exchange Project, and the Africa 2000 program. Harcourt's departure to lead the official leftist New Democratic opposition in the province, the October 1986 election of a new (rightist Social Credit) provincial government, and the November 1986 election of a business-backed Vancouver mayor, Gordon Campbell, all paved the way for the emergence of the strategic policy phase (see Smith, 1994).

The Strategic Policy Phase: 1987 to 1990

The strategic policy phase placed even more emphasis than the rational phase on establishing business and economic linkages as the central focus of international activity. To achieve this objective, Mayor Campbell established a "quiet moratorium" on additional sister city links and later began to identify possible "strategic cities." The Campbell government also upgraded the importance of economic priorities in Vancouver's sister city linkages: "A sister city rationale for the 1980s must recognize that in addition to friendship, economic and cultural opportunities must be reinforced. It is vital that governmental and nongovernmental institutions coordinate their efforts to optimize their economic benefits" (interview, October 17, 1988).

To ensure this policy goal, Campbell called on the city council to create a sister city commission and five sister city citizen committees. This was followed by a major policy review that produced the strategic city program in August 1987. The key objectives of the program included flexibility (as opposed to the formal requirements of more twinnings) and stronger business links (as opposed to a mixed range of relationships, with business often a limited priority). In identifying strategic cities, Vancouver officials gave priority to foreign cities that would act as gateways to significant national and regional economies, would provide the opportunity for generating frequent economic and cultural contact between Vancouver and overseas business people and centers, and would accord with Canadian federal government priorities. All of the cities that the Vancouver Economic Development Office identified as possibly strategic were in the Pacific Rim, except for the early established twinnings with Edinburgh and Odessa (Smith, 1992).

Perhaps the most innovative dimension of the strategic city policy thrust was the idea that without any formal designation by Vancouver, there was no necessity to even inform the targeted city of its place in Vancouver's international policy. In addition to avoiding the formalities of sister city linkages, this allowed Vancouver to add other collateral cities to its strategic plan. For example, Yokohama is Vancouver's officially acknowledged link with Japan, but city officials and business leaders seldom travel to Yokohama without including economic development-cultural stops in neighboring Osaka and Tokyo on the itinerary. Although neither Osaka nor Tokyo was designated as part of the main strategic city plan, both clearly met the strategic criteria.

The 1988 strategic city program budget was $92,000 plus a one-time $200,000 city grant to Yokohama's International Exposition. This city seeding again illustrated the impact of multipliers: The city grant was matched by grants from both the federal government and the province. With this start-up, the Vancouver-Yokohama Society was able to raise an additional $1 million—a total of $1.6 million for one international event. Vancouver slashed its strategic city spending by two thirds when budgetary restraint returned at the end of the 1980s, but the negative impact of these budget cuts were minimized because many of the global links had already reached the take-off point. Although business commitments and funding became more important in view of the limited municipal expenditure, the mayor remains the most significant individual upholding the strategic city linkages, and business leaders continue to view official city involvement as essential.

Despite the importance of Vancouver's third phase, other policy dilemmas were emerging that placed the city's strategic policy in a much broader setting by the late 1980s. Although economic considerations remained central, wide-ranging concerns were now becoming integral parts of the city's constituent diplomacy. These concerns stemmed from such events as the publication of the Brundtland Report (1987), the summer 1992 meeting of the United Nations Conference on Environment and Development, and the release of local studies on preserving livability.

The Globalist Policy Phase: The 1990s

Soldatos's (1989) definition of an international city is similar to Vancouver's definition of a strategic city: a primarily economic focus, with key cultural components. The policy reality of the 1990s and beyond, however, suggests a far more diversified city international role. With global interdependence extending into a broader range of areas, citizens are expecting their cities to join with more senior governments in meeting the new challenges. Although the strategic city components (i.e., those aimed at promoting trade and foreign investment) will continue to be central, major cities are also becoming more involved in issues related to world peace and disarmament, foreign aid, and the environment. To this point, no city in North America has developed a fully coherent *globalist* policy stance. The experience of Canadian and American cities such as Vancouver and Seattle, however, suggest that the emerging globalist policy phase will include world peace and disarmament, foreign aid, and global ecological components.

A World Peace-Disarmament Component

The experience of World War II provided much of the impetus for the earliest city-based twinnings. In Europe, Canada, Japan, and the United States, people-based exchanges to improve international understanding and promote world peace began in the decade following 1945. Although municipal paradiplomacy became more business oriented in the 1970s, some attention to the global peace issues continued. In the 1980s, there was a revival of interest in these peace-disarmament issues. The 1983 designation of Vancouver as a nuclear weapons free zone, the city's longstanding support for the Canadian peace movement, and the 1986 designation of Vancouver as a United Nations "Messenger of Peace" city on "the principle that cities represent local powers in the service of peace" all reflected this aspect of the emerging globalist policy phase. Although Mayor Harcourt's New Democratic Party ties played a major part in upgrading the importance of peace-disarmament issues, this emphasis has continued in Vancouver despite a shift in municipal administrations.

A Foreign Aid Component

Vancouver's first twinning with a sister city in 1944 was linked with the provision of humanitarian aid to war-ravaged Odessa. In the 1980s, the Harcourt government took important initiatives within the FCM for federally funded, city-based aid through the Municipal Professional Exchange and other programs. The Africa 2000 assistance of the municipality of Saanich, British Columbia, to Zomba, Malawi, is another good example of city-based foreign aid activities. (For a discussion of the Saanich, British Columbia, program, see Smith & Cohn, 1990.)

A Global Ecological Component

Vancouver became involved with ecological issues at an early stage through its long-standing commitment to preservation of regional farmland and its debate over and defeat of freeways in the 1960s. In 1991, Greater Vancouver published a review of its Livable Region Plan, *Creating Our Future,* under the chairmanship of Mayor Gordon Campbell. The review emphasized such policy concerns as air and water quality, waste treatment and disposal, and preservation of green space and arable land; this indicated that the city recognized its role in confronting such global issues as global warming (Greater Vancouver Regional District, 1990). Other cities such as Toronto are members of the international Urban Carbon Dioxide project, which is committed to developing a global plan for dealing with carbon dioxide gases. The project is based on the premise that cities can control carbon dioxide emissions and global warming through their powers to provide or deny transportation and building approvals and to develop zoning and land use regulations ("Cities Join to Fight Global Warming," 1991). More important, here as in other policy fields, cities are no longer prepared to wait for senior governments to act. The global city links developed during 50 years are being applied to policy initiatives on recycling, automobile-alternative transportation, newer waste disposal forms, and so forth. In Vancouver, aspects of each of these global components currently exist. In October 1990, for example, the city council released "Clouds of Change," a report containing 35 recommendations on atmospheric change. Although many of the proposals were outside the formal jurisdiction of the city, their inclusion indicated Vancouver's determination to deal with air quality and other environmental problems within the region (City of Vancouver, 1990).

Seattle Policy Phases

The Ad Hoc Policy Phase: The 1950s to the 1970s

Seattle's formal international activities commenced in 1951, when the city helped organize the first Japanese-American Conference of Mayors and Chambers of Commerce Presidents. Now held biennially, this conference has sought to strengthen Japanese and American (particularly West Coast) city ties. In 1957, Seattle twinned itself with the Japanese port city of Kobe; this was a response to President Dwight D. Eisenhower's 1956 "People to People" policy and also was a reflection of Seattle's significant Japanese and Asian population. In 1963, Seattle became the first U.S. city to twin with a Soviet city, the Uzbekistan capital of Tashkent in Central Asia. This initiative was indicative of Seattle's outward-looking, liberal traditions, which were not always in keeping with senior governmental policy choices. The city's active Peace Committee was a major mover in this "Great Power" city-to-city initiative, and the ongoing involvement of local Project Ploughshare members helped construct the city's Seattle-Tashkent Peace Park.[9] In 1967, the Norwegian port of Bergen became a sister city, an outgrowth of Seattle having one of the largest Scandinavian populations in the United States. Seattle received the Sister Cities International "Best Program" award for its Bergen twinning in 1970, the same year that the city elected as mayor Wes Uhlmann, a Scandinavian-American. In 1977, Seattle twinned with the Israeli city of Beersheba. As with the case of Vancouver, Seattle's early sister city links were all based primarily on traditional "people to people" cultural and educational exchanges.

The Rational Policy Phase

In 1978, Charles Royer, a liberal Democrat, replaced Wes Uhlmann as Seattle's mayor. As an active participant in the National League of American Cities and an important opponent of the Reaganomic presidency, Seattle's international activities under Royer ranged from peace-based and humanitarian concerns to tourism and business promotion. Royer was mayor of Seattle for three 4-year terms, and he therefore oversaw major changes in the city's constituent diplomacy during a lengthy period. The Seattle policy phases under Royer did not progress as systematically as in the case of Vancouver, but aspects of rational, strategic, and globalist phases are nevertheless identifiable. (As discussed, there was also some overlapping of phases in the Vancouver case.) Although rational, strategic, and globalist components were all evident in Seattle's paradiplomacy from 1978 to the present, these phases are discussed separately for purposes of comparison with Vancouver.

The major contributions of Mayor Royer's municipal government in the rational phase included a significant expansion of the city's international paradiplomacy and greater attention to organization and funding. From the small base of four sister cities, Royer developed 10 additional city international linkages by 1989: Mazatlán, Mexico (1979); Nantes, France (1980); Christchurch, New Zealand (1981); Mombasa, Kenya (1981); Chongqing, China (1983); Limbe, Cameroon (1984); Managua, Nicaragua (1984); Galway, Ireland (1986); Reykjavik, Iceland (1986); and Taejon, Korea (1989). Among the causes of this dramatic increase in the number of sister cities were the influence of local cultural communities and politically active local groups, the interest in increasing cross-border business contacts, and the priorities of an internationally oriented mayor. To coordinate the increased activity, Royer created an Office of International Affairs within his executive office. Its 1988 budget was about $240,000, with a director, a trade representative, a sister cities coordinator, a federally funded Pearson Fellow, and a secretary (interview, Keith Orton, director, Office of International Affairs, City of Seattle, July 25, 1993).

The Strategic Policy Phase

Several of the early twinnings under Mayor Royer began to include a more obvious economic dimension in the strategic policy phase. The 1986 creation of the Office of International Affairs also marked a shift from the traditional cultural-educational aspects of Seattle's paradiplomacy to a greater emphasis on international business and trade objectives. As in the case of Vancouver, Seattle's Pacific location had a major impact on the geographic focus of its international policy in the strategic phase. This strategic policy and Pacific Rim emphasis has extended into the term of Democratic Mayor Norm Rice, first elected in 1990. Thus, Seattle added the sister cities of Kaohsiung (Taiwan) and Cebu (Philippines) in 1991 and of Surabaya (Indonesia) in 1992. Important factors in the establishment of these new sister cities included the existence of important business and cultural links in Seattle and assistance from the Trade Development Alliance, which was created in early 1991. For example, the Surabaya link commenced with an invitation from the provincial governor of East Java through the U.S. consul general in the East Java capital. Seattle drew on citizens who had links with or an interest in Indonesia and created a

Seattle-Surabaya Sister City association, with the expectation that trade opportunities and cultural exchanges would develop (interview, Tsering Ottuk, Office of International Affairs, City of Seattle, February 25, 1992).

The Trade Development Alliance was created with roughly equal funding of $75,000 to $80,000 (U.S.) each from the city of Seattle, King County, the Port of Seattle (a state agency), and the Chamber of Commerce. Individual business memberships of $100 also added about $15,000 to its 1992 budget of $346,000. Rather than supplant the city's sister city activities, the alliance (with a staff of five, under the directorship of the former director of intergovernmental relations for Seattle) seeks to examine and strengthen the city-county trade connections. The alliance has completed a study of Greater Seattle-King County's commercial relationships to determine whether they are markedly different from the state as a whole (they are not) and has sought to build on the region's strong European connections. (Direct air links from Seattle to Europe are the same distance as to Tokyo, and the ethnic makeup of Greater Seattle contains significant communities with strong European backgrounds.) Seattle's sister city program has links to Norway, France, and Ireland; Perugia, Italy; Pecs, Hungary; and Gydnia, Poland. The alliance has been keen to expand its European trade opportunities in the context of Europe 1992 and the significant changes in Eastern Europe. With its strategic emphasis, the alliance has sought to identify and focus on the top target markets, recognizing that there are limits to what any city-region can do. The metropolitan region's markets reflect a Pacific Rim as well as European focus: Japan is the top trading partner, with Korea, China-Hong Kong, Taiwan, and Australia in the top 10; other strategic markets for Seattle-King County are the United Kingdom, Germany, France, and the Netherlands. Canada also ranks second or third among the Seattle region's major trading partners (interview, T. Ottuk, Office of International Affairs, City of Seattle, February 25, 1992; information also provided by the Trade Development Alliance, February 1992).

Another indication of the strategic policy phase was the passage of Resolution 28362, establishing a sister city affiliation policy for Seattle on May 13, 1991. Seattle's direct contribution is limited to $900 annually to each sister city organization, and any additional funding for sister city activities is community based. It is easier for large organizations such as the Seattle-Tashkent Association with a membership of 400 to 500 to acquire additional funding than it is for smaller associations that have as few as 20 members. The affiliation policy now seeks to require a minimum membership of 30 and to establish specific criteria for formal city affiliation, including a proposed 12-month budget and comprehensive work plan. (At least one sister city association— with Izmir, Turkey—is not city recognized, although it was established in 1988.) These criteria provide some basis for limiting the extension of formal recognition by the city. They also recognize the increased significance of economic considerations: Any new sister city affiliation with Seattle must "be the center of significant educational, cultural or political resources which offer Seattle's citizens significant exchange opportunities," or "be or have the prospect of being a major trading partner or [have] some other major economic aspect that is similar to or complementary to the economic character of Seattle" (City of Seattle, 1991, Res. 28363). The policy also restricts Seattle twinnings to cities with no other U.S. city affiliations or to cities in countries in which Seattle has no existing sister city links.

As with Vancouver, economic difficulties in the late 1980s and early 1990s have forced the City of Seattle to limit its own direct expenditures for its sister city program. The 1992 international affairs budget was set at $186,000 (U.S.), a decrease of $44,000 from 1988 levels (the total Seattle city budget proposed for 1992 was $1,309,247,602). At the same time, the Office of International Affairs became a division of the Office of Intergovernmental Relations within the Mayor's Office. Its staff includes a director, a community program coordinator, and an administrative assistant shared with the intergovernmental relations program; this is about half of the staff size reached at its high point under Mayor Royer. The downsizing, however, reflects more the economic realities of post-Reaganomic cities in America than a declining city commitment to its international programs (interview, T. Ottuk, Office of International Affairs, City of Seattle, February 25, 1992).

The Globalist Policy Phase

Seattle's globalist emphasis in its international activities emerged at a relatively early period, largely because of dissatisfaction with conservative and defense-oriented leanings of the U.S. federal government. As Kincaid (1989) notes, city council resolutions and local referenda in many U.S. cities in the 1970s and 1980s usually endorsed liberal positions on such issues as nuclear weapons testing, the divestment of funding in South Africa, and the establishment of ties with the Sandinista government in Nicaragua. Indeed, about 150 U.S. cities declared themselves nuclear weapons free zones, and more than 80 cities established sister city relationships with Nicaraguan communities, largely as a reaction to U.S. federal policies. The liberal-leaning positions were one way in which municipal governments could express dissatisfaction with the federal government's huge outlays for military activities, leaving little funding available for relieving the serious economic and social problems confronting major U.S. urban areas.

Charles Royer as a liberal Democratic mayor, therefore, felt fully comfortable in joining with other cities to oppose the policies of President Ronald Reagan toward Central America. Under Royer, Seattle designated itself a "City of Sanity" and created a citizens' commission on Central America. In 1984, in a municipal digital salute to the Reagan White House, the city made Managua, Nicaragua, a sister city. Whether it was a means of confronting a presidency that had cut urban programs dramatically or not, it did represent a city initiative "totally based on politics" (interview, City of Seattle staff, February, 1992). The sister city linkage provided a formal basis for "frequent visits to Nicaragua enabling citizens from all walks of life to see schools, churches, hospitals, and other areas of interest, and to experience first-hand the inspiration and love of the people of Managua" (City of Seattle, 1984b, p. 1). Although reflective of a strong local sentiment in opposition to Reagan's foreign policy, Seattle's Office of International Affairs took the position that foreign policy nevertheless remained the preserve of the federal government:

> The role of local government at the international level is clearly distinguished from the role of the federal government. National security, national foreign policy . . . and related matters . . . are outside the jurisdiction of the City of Seattle. . . . The educational benefits and enhanced international awareness in the local community resulting from a sister city program and other international exchanges are . . . a direct responsibility of city government. These distinctions between city and federal responsibilities at the international level seem well enough defined at the present time. (quoted in Cohn, Merrifield, & Smith, 1989, p. 97)

Despite this assertion by the Office of International Affairs, some influential Seattle groups felt that the Citizens' Central American Commission had missed this sister city distinction and was straying significantly into national foreign policy making. The ensuing struggle resulted in an initiative placed on the ballot in late 1986. This initiative called for the disbanding of the Central American commission on the grounds that the city "should not meddle in foreign relations" (City of Seattle, 1986, Initiative 30). On December 8, 1986, the initiative passed and the commission was dissolved. The Managua link remained, however, and was largely justified on the basis that such exchanges are "people to people" rather than government to government. This rationale was also used to establish and maintain Seattle twinnings with communities in both China (1983) and Taiwan (1991).

Although Seattle citizens successfully challenged the city's Central American commission, the city has actively pursued globalist policies in many other instances. For example, Seattle's local peace movement was an important participant in the city's sister city linkages with Tashkent. Seattle's Mazatlán Sister City Association adopted three orphanages in the Mexican city, raising funds for clothing, medicine, food, and toys and providing vocational training for older orphanage children. Seattle sent books and supplies to Mombasa, Kenya, as part of their sister city exchange (it also provided magazine subscriptions for the Mombasa library, hosted Crossroads Africa health professionals, and donated an emergency aid car to the African city). Seattle's Limbe, Cameroon, city link provided aid such as pharmaceuticals and a medical aid van for the Limbe Clinic in 1988 (see City of Seattle, 1984a, 1984c, 1984d). As in the case of Vancouver, Seattle has also been active on the environmental policy front. For example, Seattle has developed newer bus technology, encouraged recycling, and imposed limits on one-passenger private automobile users.

■ Conclusion

Vancouver and Seattle are both actively involved as global cities in constituent diplomacy, and their international objectives have become more broad ranging through time. This chapter has examined the growing international involvement of these two Pacific Northwest cities in four policy phases: the ad hoc, rational, strategic, and globalist phases. We will now present some conclusions regarding the determinants of municipal internationalism in each of the four phases.

As discussed, the ad hoc phase was oriented primarily to cultural and educational exchanges. A major determinant of Vancouver and Seattle's early sister city twinning in this phase was the desire to promote international understanding in the postwar period. The interest of particular ethnic groups in each city was also a major determinant of which foreign cities were selected for twinning purposes. Furthermore, the American and Canadian federal governments and other groups at the federal level often encouraged cities to begin the twinning process; for example, Vancouver's ad hoc policy phase was fully consistent with the objectives of the Federation of Canadian Municipalities for sister cities.

The rational policy phase emphasized the development of an institutional capacity for international activities and a greater focus on economic linkages. Seattle and Vancouver both increased their budgets for international purposes during this phase. Furthermore, Seattle Mayor Royer created an Office of International Affairs, and Vancouver Mayor Harcourt centered day-to-day staff responsibilities for international activities in the city's Economic Development Office. The institutionalization of policy making and the shift to a more economic focus resulted largely from the growth of interdependence in trade, foreign investment, and other economic areas. Global economic relations were having a greater impact on the growth and vitality of the Vancouver and Seattle economies, and it was therefore necessary for both cities to move beyond ad hoc and incremental policy making in this area.

The location of Vancouver and Seattle in the Pacific Northwest was also a major reason for the activism of both cities during the rational phase. At considerable distance from their respective capitals in Ottawa and Washington, D.C., the two cities sometimes have interests that diverge rather widely from those of their federal governments. For example, both cities place relatively greater emphasis on promoting Asia-Pacific ties because of their ethnic composition, geographic location, and trade and investment patterns. Mayors Harcourt and Royer were also both somewhat leftist on the political spectrum, and their international activism resulted partly from ideological differences with more conservative federal governments in Ottawa and Washington, D.C. Nevertheless, mayors in the two cities often cooperate with national leaders with different ideological orientations when it serves the interests of both levels of government. The Harcourt government, for example, worked closely with the Conservative Mulroney government in Ottawa to establish Vancouver as an international banking center.

The strategic policy phase was in fact an extension of the rational phase, with an even greater emphasis on a business-oriented economic development approach. In this phase, Vancouver shifted from adding more formal sister cities to establishing less formal linkages with strategic cities identified as gateways to important regional and national economies. Seattle adopted a somewhat different policy because it significantly expanded its formal sister city twinnings. The further growth of economic interdependence, combined with the experience Seattle and Vancouver had gained with paradiplomacy in the rational phase, were major determinants of their shift to a greater business-economic orientation in the strategic phase. Geographic location also continued to be a major determinant because Vancouver's strategic cities and Seattle's newer sister cities were concentrated primarily in the Pacific Rim. Furthermore, the ethnic composition of both cities was important in the strategic phase, as it had been in the ad hoc and rational phases; this helps to explain, for example, the two cities' emphasis on establishing economic linkages with European as well as Pacific Rim cities.

The globalist phase adds significantly to the policy content of the previous phases by also including issues such as world peace and disarmament, international aid, and environmental sustainability. Despite the expanding scope of municipal internationalism, the establishment of business and economic linkages probably continues to be the most important paradiplomatic preoccupation of Vancouver and Seattle. The two cities' policy shifts during the globalist phase indicate a recognition that global interdependence has significant local effects in an ever widening array of issue areas. Thus, cities are adversely affected not only by economic downturns: Environmental disasters such as major oil spills, military conflicts that could involve the use of nuclear weapons, and serious problems in the Third World affecting the flow of refugees and immigrants can all have a major effect on cities such as Seattle and Vancouver. The failure of national governments to deal with the growing array of intermestic issues has moved local governments and citizens to increase their global consciousness in the globalist phase.

Although this chapter has focused specifically on municipal internationalism, we should note that Vancouver and Seattle's activities form only one part of a developing constituent diplomacy in the Pacific Northwest. British Columbia and Washington State are the central motivating forces in these growing microregional, macroregional, and global interactions, but Vancouver and Seattle as the major cities in the microregion also have a significant role. These initiatives have characteristics of all four phases of municipal paradiplomacy discussed in this chapter. Examples of established and developing microregional initiatives include the British Columbia-United States Oil Spill Task Force (linking Alaska, British Columbia, Washington State, Oregon, and subsequently, California); the Pacific Northwest Economic Region (PNWER); the Georgia Basin (British Columbia)/Puget Sound (Washington State) Initiative; the Cascade Corridor Commission (an advisory body of governments in Canada, the United States, British Columbia, Washington State, and Oregon); and other Cascadia activities involving government and/or private sector participation. (For further discussion of international activity in the Pacific Northwest, see Artibise & Hill, 1993; Cohn & Smith, 1993; Smith, 1993; Smith & Goddard, 1993.) As global cities, Vancouver and Seattle sometimes act independently of other regional initiatives. These two cities, however, also form a major part of the developing global region in the Pacific Northwest.

NOTES

1. The British Columbia Round Table on the Environment and the Economy, established by the provincial government with municipal involvement in 1990, has produced a series of studies on the Georgia Basin (and Puget Sound) metropolis. See, for example, B.C. Round Table's (1993) *Georgia Basin Initiative: Creating a Sustainable Future.* See also Artibise and Hill (1993), Davis and Hutton (1992), Gunton (1991), and Horne and Robson (1993).

2. The terms *ingressive* and *egressive* are taken from Duchacek (1986). Duchacek, however, does not define them consistently (see pp. 208-209 of his study). *Webster's New International Dictionary* (1961) defines *egress* as "to go out, to issue forth" (p. 821) and *ingress* as the "act of entering; entrance" (p. 1278).

3. Cooperation and conflict often coexist in a relationship. Furthermore, some competitive situations are relatively friendly, whereas others are more closely associated with conflict. For a detailed examination of the relationship between these three variables, see Cohn (1990).

4. For an in-depth analysis of this typology applied to cities, see Cohn and Smith (1990) and Smith and Cohn (1994). For an analysis of this typology applied to provinces and states, see Cohn and Smith (1993).

5. Written response to the New International Cities Era (NICE) questionnaire from Seattle's Office of International Affairs staff, 1988.

6. On the discrepancy between capacity and formal authority in federations such as Canada, see Smith (1988).

7. The material in this section is taken primarily from Cohn et al. (1989), "Seattle Datasheet," (City of Seattle, Division of International Affairs, 1992), Trade Development Alliance of Greater Seattle (1992), and an interview with Keith Orton, director of the City of Seattle Office of Intergovernmental Affairs (July 1993).

8. The trade data in this section are taken from the British Columbia Ministry of Economic Development *British Columbia Facts and Statistics* (1975-1994); the British Columbia Ministry of Finance and Corporate Relations *British Columbia Economic and Statistical Review* (1975-1994); the British Columbia Ministry of International Business and Immigration *British Columbia: Trade 1989* (1989); British Columbia Ministry of Finance and Corporate Relations *British Columbia International Exports, 1990* (1990); British Columbia Ministry of Economic Development, Small Business and Trade *British Columbia International Exports, 1991* (1991); British Columbia Ministry of Finance and Corporate Relations (1993); City of Vancouver Economic Development Office (1993); and interview with Don White, policy adviser, Trade Policy, International and Intergovernmental Branch, Ministry of Economic Development, Small Business and Trade, British Columbia (April 1993).

9. Project Ploughshare is a nongovernmental international aid agency.

REFERENCES

Alger, C. F. (1987). Linking town, countryside and legislature to the world. *International Studies Notes, 13*(3), 57-63.

Artibise, A. F. J., & Hill, J. (1993). *Governance and sustainability in the Georgia Basin.* Victoria: British Columbia Round Table on the Environment and the Economy.

British Columbia Ministry of Economic Development. (1975-1994). *British Columbia facts and statistics.* Victoria: Author.

British Columbia Ministry of Economic Development, Small Business and Trade. (1991). *British Columbia international exports, 1991.* Victoria: Author.

British Columbia Ministry of Finance and Corporate Relations. (1975-1994). *British Columbia economic and statistical review.* Victoria: Author.

British Columbia Ministry of Finance and Corporate Relations. (1990). *British Columbia international exports, 1990.* Victoria: Author.

British Columbia Ministry of Finance and Corporate Relations. (1993, March). *BC Stats.* Victoria: Author.

British Columbia Ministry of International Business and Immigration. (1989). *British Columbia: Trade 1989.* Victoria: Author.

British Columbia Round Table on the Environment and the Economy. (1993). *Georgia Basin Initiative: Creating a sustainable future.* Victoria: Author.

Brundtland, G. H. (1987). *Our common future.* New York: United Nations, World Commission on Environment and Development.

Center for Policy Alternatives. (1993, July). *Summary of state and municipal actions relating to the North American Free Trade Agreement (NAFTA)* (3rd ed.). Washington, DC: Author.

Cities join to fight global warming. (1991, June 13). *Toronto Star,* p. A10.

City of Seattle. (1984a). *Passport to sister cities: Limbe.* Seattle, WA: Author.

City of Seattle. (1984b). *Passport to sister cities: Managua.* Seattle, WA: Author.

City of Seattle. (1984c). *Passport to sister cities: Mazatlán.* Seattle, WA: Author.

City of Seattle. (1984d). *Passport to sister cities: Mombasa.* Seattle, WA: Author.

City of Seattle. (1986, December 8). *Initiative 30.* Seattle, WA: Author.

City of Seattle. (1991, May 13). *Resolution 28363: Sister City Affiliation Policy.* Seattle, WA: Author.

City of Seattle, Division of International Affairs. (1992, January). *Seattle datasheet.* Seattle, WA: Author.

City of Seattle, Office of International Affairs. (1988). [Written response to New International Cities Era (NICE) questionnaire]. Seattle, WA: Author.

City of Vancouver. (1990). *Clouds of change.* Vancouver, BC: Author.

City of Vancouver, Economic Development Office. (1993, April). *Vancouver economic data base.* Vancouver, BC: Author.

Cohn, T. H. (1990). *The international politics of agricultural trade: Canadian-American relations in a global agricultural context.* Vancouver: University of British Columbia Press.

Cohn, T. H. (1993). *The intersection of domestic and foreign policy in the NAFTA agricultural negotiations* (Canadian-American Public Policy Series, Monograph No. 14). Orono: University of Maine.

Cohn, T. H., Merrifield, D. E., & Smith, P. J. (1989). North American cities in an interdependent world: Vancouver and Seattle as international cities. In E. H. Fry, L. H.

Radebaugh, & P. Soldatos (Eds.), *The new international cities era: The global activities of North American municipal governments* (pp. 76-79). Provo, UT: Brigham Young University, David M. Kennedy Center for International Studies.

Cohn, T. H., and Smith, P. J. (1990). A typology of city-based international involvement: The case of Vancouver. In *Processus d'internationalization des villes*, conference proceedings. Lyon, France: Groupe Nice, MRASH.

Cohn, T. H., & Smith, P. J. (1993, May). *Constituent diplomacy policy determinants in British Columbia: Developing a global region in the Pacific Northwest.* Paper presented at the Colloquium on Les Déterminants de la Politique International des Provinces Canadiennes, Université Laval, PQ.

Davis, C., & Hutton, T. (1992). *The role of services in metropolitan economic growth.* Victoria: British Columbia Round Table on the Environment and the Economy.

Duchacek, I. D. (1986). *The territorial dimension of politics: Within, among, and across nations.* Boulder, CO: Westview.

Duncan, S. (1992, June). *Investment performance in Canada: Understanding the big picture* [*Canada 2000* pamphlet]. Calgary, AB: Canada West Foundation.

Fagan, D. (1993, September 7). NAFTA deals may exempt Canada: Provinces must ratify measures that apply to their jurisdiction. *Globe and Mail,* p. B8.

Federation of Canadian Municipalities. (1988). *A practical guide to twinning.* Ottawa, ON: Author.

Fry, E. H. (1986). The economic competitiveness of the western states and provinces: The international dimension. *American Review of Canadian Studies, 16*(3), 301-312.

General Agreement on Tariffs and Trade. (1986, July). *Text of the General Agreement.* Geneva, Switzerland: Author.

Goldrick, M. D. (forthcoming). The impact of global finance in urban structural change: The international banking centre controversy. In J. Caulfield & L. Peake (Eds.), *Cities and citizens: Critical research and Canadian urbanism.* Toronto, ON: University of Toronto Press.

Greater Vancouver Regional District. (1990, September). *Creating our future: Steps to a more livable region.* Burnaby, BC: Author.

Gunton, T. (1991). *Economic evaluation of environmental policy.* Victoria: British Columbia Round Table on the Environment and the Economy.

Hocking, B. (1986). Regional governments and international affairs: Foreign policy problems or deviant behavior. *International Journal, 41,* 477-506.

Horne, G., & Robson, L. (1993). *British Columbia: Community economic dependencies.* Victoria: British Columbia Round Table on the Environment and the Economy.

Johannson, P. R. (1978). Provincial international activities. *International Journal, 33,* 357-378.

Jones, V. (1988). Beavers and cats: Federal-local relations in the United States and Canada. In H. P. Oberlander & H. Symonds (Eds.), *Meech Lake: From centre to periphery* (pp. 88-126). Vancouver: University of British Columbia Centre for Human Settlements.

Keohane, R. O., & Nye, J. S., Jr. (1976). Introduction: The complex politics of Canadian-American interdependence. In A. B. Fox, A. O. Hero, Jr., & J. S. Nye, Jr. (Eds.), *Canada and the United States: Transnational and transgovernmental relations* (pp. 3-15). New York: Columbia University Press.

Kincaid, J. (1989). Rain clouds over municipal diplomacy: Dimensions and possible sources of negative public opinion. In E. H. Fry, L. H. Radebaugh, & P. Soldatos (Eds.), *The new international cities era: The global activities of North American municipal governments* (pp. 240-246). Provo, UT: Brigham Young University, David M. Kennedy Center for International Studies.

Kincaid, J. (1990). Constituent diplomacy in federal polities and the nation-state: Conflict and co-operation. In H. J. Michelmann & P. Soldatos (Eds.), *Federalism and international relations: The role of subnational units* (pp. 54-75). New York: Oxford University Press.

Krasner, S. D. (1983). Structural causes and regime consequences: Regimes as intervening variables. In S. D. Krasner (Ed.), *International regimes* (pp. 1-21). Ithaca, NY: Cornell University Press.

Manning, B. (1977). The Congress, the executive and intermestic affairs: Three proposals. *Foreign Affairs, 55,* 306-324.

North American Free Trade Agreement Between the Government of Canada, the Government of the United Mexican States and the Government of the United States of America. (1992, December).

Oberlander, P., & Smith, P. J. (1993). Governing metropolitan Vancouver: Regional intergovernmental relations in British Columbia. In D. Rothblatt & A. Sancton (Eds.), *Metropolitan governance: American/Canadian intergovernmental perspectives* (pp. 329-373). Berkeley: University of California, Institute of Governmental Studies.

Rosenau, J. N. (1990). *Turbulence in world politics: A theory of change and continuity.* Princeton, NJ: Princeton University Press.

Ruggie, J. G. (1983). International regimes, transactions, and change: Embedded liberalism in the postwar economic order. In S. D. Krasner (Ed.), *International regimes* (pp. 195-231). Ithaca, NY: Cornell University Press.

Seelig, M. Y., & Artibise, A. F. J. (1991). *From desolation to hope: The Pacific Fraser region in 2010.* Vancouver, BC: Board of Trade.

Smith, P. J. (1986). Regional governance in British Columbia. *Planning and Administration, 13*(Autumn), 1-20.

Smith, P. J. (1988). Local-federal government relations: Canadian perspectives, American comparisons. In H. P. Oberlander & H. Symonds (Eds.), *Meech Lake: From centre to periphery* (pp. 127-138). Vancouver: University of British Columbia Centre for Human Settlements.

Smith, P. J. (1992). The making of a global city: Fifty years of constituent diplomacy: The case of Vancouver. *Canadian Journal of Urban Research, 1,* 90-112.

Smith, P. J. (1993). Policy phases, subnational foreign relations and constituent diplomacy in the United States and Canada: City, provincial and state global activity in British Columbia and Washington. In B. Hocking & M. Forsyth (Eds.), *Foreign relations and federal states* (pp. 211-235). London: Leicester University Press.

Smith, P. J. (1994). Labour markets and neo-conservative policy in British Columbia, 1986-1991. In A. Johnson, S. McBride, & P. J. Smith (Eds.), *Continuities and discontinuities: The political economy of social welfare and labour market policy making in Canada* (pp. 291-305). Toronto, ON: University of Toronto Press.

Smith, P. J., & Cohn, T. H. (1990, May). *Municipal and provincial paradiplomacy and intermestic relations: B.C. cases.* Paper presented at the annual meeting of the Canadian Political Science Association, University of Victoria, Victoria, BC.

Smith, P. J., & Cohn, T. H. (1994). International cities and municipal paradiplomacy: A typology for assessing the changing Vancouver metropolis. In F. Frisken (Ed.), *The changing Canadian metropolis: A public policy perspective* (Vol. 2, pp. 613-655).

Berkeley: University of California, Institute of Governmental Studies; Toronto, ON: Canadian Urban Institute.

Smith, P. J., & Goddard, A. M. (1993, November). *The development of subnational foreign relations: The case of the Canada-U.S. Pacific Northwest.* Paper presented at the biennial conference of the Association of Canadian Studies in the United States, New Orleans, LA.

Soldatos, P. (1989). Atlanta and Boston in the new international cities era: Does age matter? In E. H. Fry, L. H. Radebaugh, & P. Soldatos (Eds.), *The new international cities era: The global activities of North American municipal governments* (pp. 37-72). Provo, UT: Brigham Young University, David M. Kennedy Center for International Studies.

Soldatos, P. (1990). An explanatory framework for the study of federated states as foreign-policy actors. In H. J. Michelmann & P. Soldatos (Eds.), *Federalism and international relations: The role of subnational units* (pp. 34-53). New York: Oxford University Press.

Statistics Canada. (1988). Corporations and Labour Unions Return Act, Part I: Corporations. In *Annual Report* (p. 70). Ottawa, ON: Author.

Thorne, P. M. (1993). *1992/1993 British Columbia inbound investment study.* Vancouver, BC: Author.

Trade Development Alliance of Greater Seattle. (1992, June). *International market report: Greater Seattle/King County's current and opportunity foreign markets.* Seattle, WA: Author.

United Nations. (1986). *Messenger of peace designation: City of Vancouver.* Vancouver, BC: Author.

Webster's new international dictionary of the English language (2nd ed.). (1961). Springfield, MA: Merriam.

Conclusion and Epilogue: The Future of Cities and Their Policies in the Global Economy

GARY GAPPERT

Introduction

This book was developed with a concern for the emergence of cities as important actors in the global and international economy. Traditionally, the international economy focused on economic relations between nations. Today the international set of economic relations has become truly globalized, and nations are no longer the only significant players. Cities on every continent are now major actors in global economic development.

Globally engaged cities throughout North America are strategically planning their own economic development activities and establishing relations and functional linkages with other cities in the dramatically altered space of a global economy. With the passage of the General Agreement on Tariffs and Trade (GATT) on December 1, 1994, and the implementation of the North American Free Trade Agreement (NAFTA), dramatic changes are confronting both nations and cities in the management of their economic growth, development, and relationships. New intercity linkages on a global scale consist of experience sharing; joint lobbying for issues of common interest; establishment of linkages between educational, research, and cultural institutions; and a variety of related activities. Most major cities today recognize the need to implement policies that will enhance their international competitiveness.

This book is organized around four major themes or topics:

1. The dynamics of urban international engagement in the global economy
2. Municipal policy issues and relationships and globalization

3. Municipal networking and cross-border intergovernmental cooperation
4. Case study perspectives in North America

In addition, the principle founder of urban economics, Wilbur Thompson (1968), has provided an introduction in which he updates his original *A Preface to Urban Economics* with perceptions of the role of cities in the emerging global economy.

The impacts of the global economy on the internationalization of cities is the subject of inquiry that provides the unifying theme of this volume.[1] This epilogue reviews some of the more salient conclusions and projections so that the reader can gain insights both into this new area of research and into the policy and planning needs of cities concerned with their future in the global society (see Knight & Gappert, 1989).

Major Themes

The writers in this volume are primarily economists, planners, and political scientists with an orientation toward political economy or intergovernmental relations. Most are active at some level in municipal affairs and policy development. This section will consider some of their conclusions and questions about policy choices for the future.

The Dynamics of Urban International Engagement in the Global Economy

The world is experiencing the evolution of a new form of international relations that is more geoeconomics than geopolitics. New forms of international and transnational engagement are developing. In Chapter 1, Fry reviews the growth of the international relations and the paradiplomatic activities of cities in recent years. He cites the term *intermestic politics* to reflect the growing overlap of *inter*national and do*mestic* issues to capture the "essence of new challenges facing these subnational governments in North America." Fry reviews these efforts and analyzes their effectiveness. He concludes with 10 recommendations for municipal leaders for whom "the quest to internationalize U.S. cities should remain a top priority."

In Chapter 2, Kresl offers an important new twist on the competitiveness debate by changing the focus from the national economy to the urban economy. Kresl stresses that the concept of a city's international competitiveness is quite different from the concept of an international or global city. The latter concept has been articulated by Soldatos (1991, 1993), Knight (1989), and others and is primarily concerned with the evaluation of the degree to which a city is in fact international in its connections with the rest of the world economy. Kresl assesses the extent to which a city may dramatically increase its global economic competitiveness without increasing the degree to which it is an international city. His evaluation of the importance of various determinants of competitiveness goes beyond traditional economic issues and considers important issues of governance, urban amenities, environment, and public-private sector cooperation.

In Chapter 3, Kincaid considers the contemporary world order in which urban economies are being internationalized. He reviews the forces of globalization on a worldwide basis. For him, the primary concern is the tension between integrative forces at the global and international level and disintegrative and fragmentation forces within nations. The forces of *intra*national decentralization have a particular impact on cities and municipalities. In particular, Kincaid cites the emerging competition between "citizenship" at the municipal level and "consumership" on a global scale. With an increase in labor mobility in a global economy, interest in local civic responsibility has lessened. Kincaid also cites the potential access for foreign suppliers to the $200 billion market in state and local procurement activities following the ratification of GATT. The management challenges are enormous.

Municipal Policy Issues, Relationships, and Globalization

Urban regimes and municipal governments are making themselves into proactive agents in the global economy. This is a return to a function of city-states that predates the development of nation-states in Europe some three or four centuries ago. But the independent and interdependent economic function of cities in today's complex technological society is in a state of flux and evolution. In this section, two authors address the relationship between internationalization and the restructuring and revitalization of municipal policy in cities in North America.

In Chapter 4, Levine examines the extent to which economic change in the global marketplace has fostered inequality within cities. For some time, Levine has tracked the growth of the dual labor markets and economies in cities such as Baltimore and Montreal. As cities begin to participate more in the global economy, not all participants are equal. Municipal leaders must be aware of the growth of a "dual city" even as a city become more successful in its international competitiveness.

Hiernaux completes this part with a perspective from Mexico in Chapter 5. He explicitly emphasizes a Mexican point of view in assessing the "territorial logic" behind the new continentalism in North America in the process of globalization. He is particularly concerned with the "progressive disintegration of the so-called city system in Mexico" and the interplay between global change and urban regional territorial issues. Nicolas cites in particular "menacing poverty" and pollution as structural limitations of Fordist economic policies for cities in an otherwise globalizing economy. His analysis, in which he is concerned as much for the losers as for the winners, is borne out by the continuing crisis in Mexico.

Municipal Networking and Cross-Border Intergovernmental Cooperation

Clement, writing from the vantage point of the San Diego-Tijuana cross-border region, introduces this theme in Chapter 6 with observations on the paradigm shift with respect to the new role of cities and regions in the increasingly globalized economic environment. Clement provides a conceptual overview of the necessary elements for a municipal developmental strategy in the new global economy. He concludes his chapter with examples from his region.

Internationalization and the development of a global economy can be viewed from different perspectives. In Chapter 7, Shachar from Israel examines European world cities and reviews the evolution of the world city concept. The historical perspective attempts to relate particular types of urban development to the global economic restructuring process. The role of infrastructure upgrading and the loci of advanced business services play important roles in this restructuring. This chapter concludes with an assessment of how the European urban system is being redefined and reclassified.

In Chapter 8, Proulx demonstrates that an understanding of city-regional growth in the global economy calls for an explicit introduction of space and spatial relationships into the analysis. Using data that reflect the geographic trade of Quebec with states in the United States, his research discounts to a degree the importance of the proximity provided by border areas. Proulx indicates a dramatic "recomposition of economic space" as a result of "both increasing globalization and growing regionalization." He believes that the local-regional space of a border region is too restricted and that strategic planning for cities must consider the entire economic space provided by new forms of integration.

Case Study Perspectives in North America

In a number of regions in North America, cities and urban regions that span an international political boundary share some unique relationships, perspectives, and problems. These border regions provide a fascinating research opportunity for assessing the evolution of new forms of functional regional cooperation between cities and municipalities.

Schmidt opens this section by presenting a view from Mexico in Chapter 9. This chapter has a strikingly different perspective. It demonstrates how the U.S.-Mexican border area, an inherently international region, "has apparently lost its opportunity to enjoy the comparative advantage of both countries." Schmidt reviews the obstacles for planning a bi-national metropolis. A key issue is that an industrial model was applied "without ecological adaptation, depleting and damaging scarce resources, a process which by now has resulted in a catastrophic border scenario with little appeal for new investors, who may find the border a hindrance instead of an advantage." This border region is the counterpoint to Cascadia.

In Chapter 10, Artibise reviews an emerging model of urban growth management in the Vancouver-Seattle-Portland corridor, known in that region as "Cascadia." The main issues that drive functional cooperation in this region are "sustainability, rapid urban growth, transportation, trade, tourism, and economic development." Growth management is a theme that unites the cities and their regions in this highly environmentally sensitive area. As Artibise demonstrates, in this region the challenges of "sustainable urbanization" are being explicitly addressed even as economic growth is also being pursued.

Artibise's regional perspective is followed by Cohn and Smith's presentation of the cases of Vancouver and Seattle in Chapter 11. They use this set of cross-border relationships to compare and contrast two efforts at becoming global cities in this important and unique "bi-national global region." They provide two global city typologies that reflect municipal governments as both a primary and a mediating actor. A four-phase policy process of becoming a global city (ad hoc, rational, strategic, and globalist) is also identified and discussed. Their chapter serves as a guide to the municipal internationalism of border cities.

Overview by Wilbur Thompson

Thompson opens his introduction with the statement that "the global society is quite rightly depicted in these pages as consumerism writ large." Global economic growth ultimately is driven by the role of consumer sovereignty. All the neoclassical international economic trade economists since Alfred Marshall have always argued that the ultimate beneficiary of free trade is the consumer.

But Thompson reminds us that for producers there are as many losers as winners. The champions of free trade all "focus on the larger side of the pie and have little to say on the size of the shares." He goes on to indicate that "the dark side of these bright facets of economic progress is that they all entail sharp changes and abrupt displacements." Many of the chapters in this book address the issues of how cities can and are organizing themselves to be winners in the wake of global restructuring. As Thompson concludes in one section, state and local governments hold the key to workable compromises, adjustments, and aggressive strategic planning in the preparation for new and different work, reflecting both the new international and *regional* division of labor in the global economy. He then applies some of his well-known urban economic theories to the policy choices that cities face today (Thompson, 1968).

Thompson leaves us with two challenges—what he calls "environmentalism in counterpoint" and the "framing of testable hypotheses." He suggests that other goals "such as environmentalism will arise to modify rampant consumerism." It is suggested that "global competition will probably produce greater divergence in natural environments." Indeed,"it is hard to write a scenario in which city *growth* does not clash with the protection of the environment." With regard to testing new hypotheses, Thompson indicates that cities are often reactive in their policies because they are timid in forecasting the future. In the rest of this epilogue, I address the challenges of environmentalism and the framing of testable hypotheses that project future behavior.

■ The Intrusion of Environmental Change

Since the early 1970s, with the publication of *The Limits to Growth: A Report for the Club of Rome* (Meadows et al., 1972), the environment has loomed large as the new doomsday specter at the global feast of industrial and urban development. The "global problematique" is no longer an apparition. As the 21st century approaches, cities exist in a time of rapid but uncertain environmental change. Industrialization and urbanization have increasingly affected the quality of urban air, water, and soil; the instability of the climate has augmented the need for worldwide urban assessment and action.

Until recently, specific urban environmental issues were addressed primarily by scientists, engineers, and narrow policy specialists in the fields of air, water, solid waste, transportation, and land use. The urban economic development perspective was held to be suspect for a quality environment.

Schmidt provides similar cautions in Chapter 9 of this volume: He writes that "regional integration will require fiscal coordination, the development of environmental brown and green agendas focusing on a new developmental philosophy, and a new concept based on prevention and sustainable development." Furthermore, to compete on the international scene:

> Urban development needs to be considered not only as a housing problem but also as a question of structure of services, including the environment, water (drinking and sewage), handling of toxic and solid waste, supplies (supermarkets, groceries, and other items), education, mail . . . disaster preparedness, law enforcement . . . pollution generated in one city but affecting its neighbor, and garbage generated in one city and disposed in the other.

The urban environmental focus—the so-called brown agenda[2]—has been specialized and somewhat unfocused. But the emergence of the "urban problematique" is beginning to displace the simplicity of the boom-or-bust, grow-or-decline paradigm that has preoccupied civic communities and regimes throughout North America and Western Europe since the mid-19th century. The new urban paradigm must be based on a balance between economy, ecology, and equity (Gappert, 1993).

Well-governed municipalities have regimes that balance an apparent dichotomy between growth policies and green policies. Lemberg (1989), a Danish planner and economist, has presented this shift in Figure 1. The new growth-green dichotomy has already been manifested extensively in municipal elections in Western Europe. Given the tradition of municipal boosterism in U.S. cities, this dichotomy has been more muted in North America. Nevertheless, its dimensions are clearly present in such bellwether cities as Toronto and Vancouver in Canada and in Seattle, Denver, and even Los Angeles in the United States.

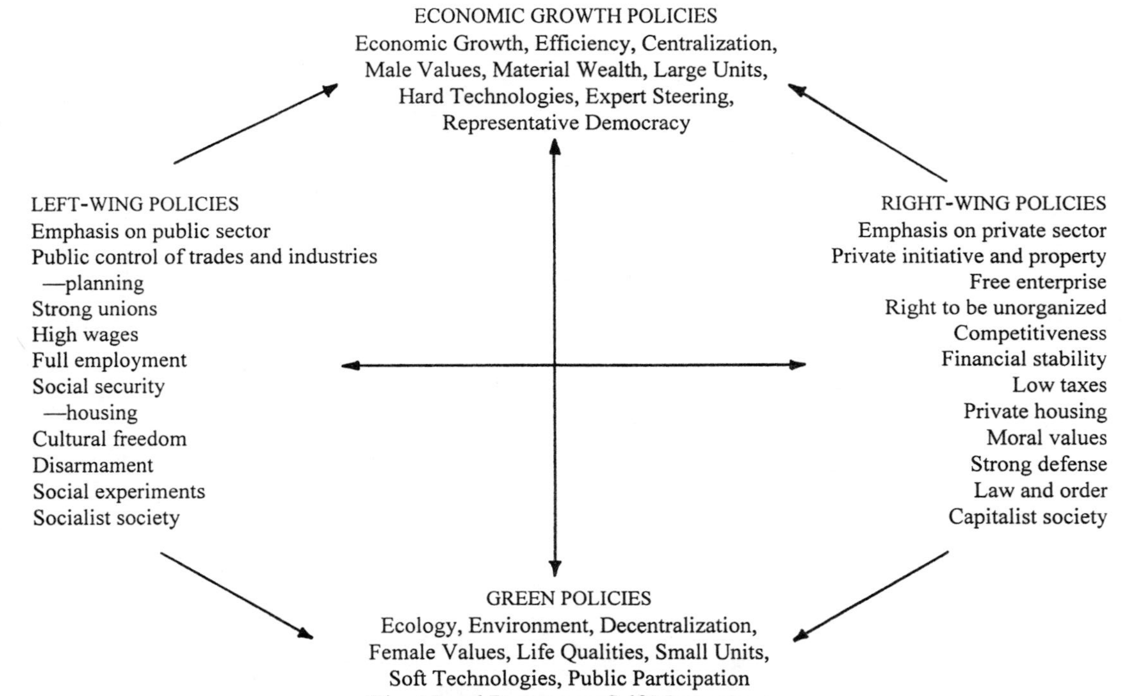

Figure 1. European Urban Paradigm Shift
SOURCE: Lemberg (1989), p. 215.

Artibise introduced the theme of sustainability in his chapter on the Pacific Northwest. The theme of sustainable development is a concept put forward by the World Commission on Environment and Development (1976) in its report *Our Common Future.* Sustainable development is a process by which business and society attempt to live in harmony with the natural environment. In the future, society will need to invest more of its intellectual and economic resources to adapt cities to their natural environment after decades of industrial abuse (Gappert, 1993).

Although a balance between economic growth policies and environmental protection will be a major preoccupation in the decade ahead in most major cities competing in the global economy, this resolution will depend on enlightened leadership and coalition trade-offs. International trade theory can be useful in beginning to formulate some simplifying propositions.

■ The Role of International Trade Theory

In projecting the future of cities and urban environments in the next century with liberalized trade regulations and a more dynamic international economy, the assumptions about international economic development and its impact on cities need to be clarified. A review of traditional neoclassical international economic trade theory (Balassa, 1981; Chenery, 1979; Hirschman, 1958; Kindleberger, 1964; Maizels, 1963) shows that many of the issues revolve around the relative mobility and immobility of the three aspects of production: land, labor, and capital. Capital was assumed to be relatively mobile, and land, of course, was quite immobile. Labor was shown to be much more irregular in its mobility.

In the last several decades, new patterns of immigration have made labor much more mobile and less place bound. New patterns of technology transfer, primarily through multinational corporations, have only increased the mobility of capital on a global scale. Paauw (1970) and others (e.g., Paauw & Fei, 1973) projected some of these consequences more than 20 years ago. Land has not become more mobile but in fact has become degraded through environmental abuse in both the development of agriculture and industry and through the relentless migration to overcrowded cities and urban regions (Brown, 1993; Gappert, 1993).

The next section attempts to respond to Thompson's challenge to generate some testable hypotheses. With the heightened mobility of capital and with an increasingly mobile labor force that can cross most international boundaries in one way or another, the environmental degradation of land and cities may have the most to do with the challenges and opportunities faced by cities in the emerging global economy of the next century.

Propositions

This section states a number of almost self-obvious economic propositions drawn from neoclassical economic theory and discusses them from a social and municipal perspective. Because most of the discussion of the benefits of trade in a global economy relates to the consumer, these propositions are focused both on producer behavior and the consequences for workers and their environment. (See Ahmal, 1978; Dorfman & Dorfman, 1972; Schumpeter, 1967; Zuvekas, 1979.)

Proposition #1: Footloose industries generally seek low-cost, low-wage, and loosely regulated (LCLWLR) locations in which to locate or to relocate their production facilities.

The enhanced mobility of capital in the global economy has accelerated the movement and relocation of plant facilities across state, regional, and national boundaries. This mobility has been advanced by dramatic changes in transportation and communications and will continue unless checked by political instability in some areas and dire poverty in others.

Proposition #2: The relocation of footloose industries in LCLWLR locations tends to raise both wages and income levels in the receiving locations and to reduce wage pressures and employment levels in the source locations.

These first two assumptions reflect trends that tend, through time (all other things being equal), to equilibrate wage and employment levels between both receiving and source locations. The eventual result, crudely stated, is that the development of a global labor market because of liberalized trade and capital mobility means that the expansion of a global working class-middle class will lead to the expansion of the underclass in the major cities of the metropolitan economies of the highly developed countries such as the Group of Seven (G-7: Britain, Canada, France, Germany, Italy, Japan, and the United States).

The trade-off to the transfer of low-wage, low-skill jobs to the periphery is that the price of an expanding Third World middle class is the growth of pockets of periphery in the metropolitan nations of the First World, especially in their cities. Also, the growth of the new global middle class assumes that the economic surplus generated in LCLWLR locations is in fact increasingly shared out to members of the urban working class, not retained at the center or corrupted away. In some countries, that is probably a shaky assumption because the top elite appropriates the surplus for itself and stores it in political safe havens. The early 1994 events in Mexico indicated the degree of popular resistance to the perceived consequences of NAFTA. On the other hand, car exports to Mexico do indicate some increase in middle-class purchasing power. Other data indicate substantial Mexican purchase of capital equipment from the United States, which supports the American working class in states such as Ohio, Illinois, and Michigan. In Ohio, the leading exports were industrial equipment and chemicals.

There are some exceptions to the expressed trade-off. Let's consider some additional propositions.

Proposition #3: The growing middle class in the receiving locations will contribute an additional propensity to consume exports from the metropoles in both consumer and producer goods and services.

Traditionally, the economic expansion in developing sites blunted some of their growing tendency to consume imports through a vigorous process of import substitution (Jacobs, 1976). But with today's global consumerism, much of the consumption also consists of services from the metropoles (Caribbean tourists to Orlando, foreign elites to the Cleveland Clinic, Mexican skiers to Colorado, "commissions" and other "fees" such as payoffs to the political safe havens of global banking facilities wherever, etc.).

It could be argued that the diversion of the economic surplus away from the receiving location mostly raises incomes away from both that urban site and the source location as the money is reinvested by the elite in political safe houses. But in North America, the increase in continental purchasing power is evident. U.S. Commerce Department data indicate that U.S. exports to Mexico increased by almost 18% in the first 10 months after NAFTA went into effect (Harbrect, 1994).

Proposition #4: Disinvestment in low-value-added industries drawn to LCLWLR locations allows for investment funds to flow to high-technology investment opportunities in the knowledge centers of the metropoles.

The work of Knight (1992) on the knowledge-based city indicates the need to transfer resources from declining sectors using low skills to sectors requiring high skills that command high wages. The so-called knowledge worker (Drucker, 1994) represents a new form of human capital investment that also requires a redirection of investment into postsecondary education of various types. In Akron, Ohio, as tire production jobs declined, there was substantial investment in a new College of Polymer Sciences at the University of Akron and similar research at Case Western Reserve in Cleveland.

Proposition #5: The further expansion of the global, free trade economy will benefit significantly those metropole cities that are control centers for corporations, financial institutions, and technological development.

This proposition is supported by research from many contributors and has been strongly championed by Knight (1989). In this proposition, the enduring comparative advantage held by cities in the developed metropolitan nations is their ability to create new productive knowledge (e.g., knowledge that contributes to productivity in advanced industries as well as advanced leisure industries).

The arguments behind the five propositions cited above are reasonably well known and have been rigorously advanced by those traditional neoclassical economists with an eye on the real world and the changing continental and global system(s) of cities. Attention to the changing comparative advantage of cities is important. The role of new knowledge classes and the role of the dislocated classes will create political tension, social anger, and the like, and the remedies will include well-intended retraining programs such as those advocated by Secretary of Labor Robert Reich. If the tension and anger increases sufficiently to cause civil unrest or territorial revolts, the pace to a liberalized global economy will be slowed, but only in rare cases will it be reversed. Enough of the economic surplus could be directed to strategic side payments to pacify significant areas of severe economic pain. Some cities will compete vigorously to direct those side payments to their own dislocation needs. Other cities will focus on developing new sources of comparative and competitive advantage. Others will do both.

The Environmental Conundrum

It is less clear what environmentalism means for cities in the global century. For low-cost, low-wage, loosely regulated sites, the central issue is whether the environmental costs will be tolerated, avoided, or ignored. A number of additional propositions need to be reviewed.

Proposition #6: The changing consumer propensities of the expanding middle class in LCLWLR sites will slowly but surely cause them to develop a preference for environmental quality.

This is a controversial proposition to which several exceptions can be made. First, it is unlikely that the first working-class/middle-class generation will value environment over jobs. Their changing propensity curve is likely to be lumpy and discontinuous. And the underemployed class will certainly prefer jobs over environment. In the aggregate, the curve will only slowly turn upward. It could even take two generations, if not longer. Nevertheless, by substituting environmental safety for environmental quality and relating it to occupational safety, a different preference curve might rise more sharply and more quickly.

Proposition #7: Producers with short-term horizons will pollute and relocate when their pollution becomes unacceptable locally.

This proposition is reflected in Chapter 9 by Schmidt in which he cites the role of over 2,000 *maquiladoras.* His analysis indicates that "the U.S.-Mexican border region faces a dilemma. It can follow the traditional industrialization model, the one that created environmental, social, and economic problems in many parts of the world. Or it can search for a new model of sustainable development, creating a brown and green agenda."

Proposition #8: Producers with longer horizons will support environmental regulation and management so far as it enhances and maintains their productive advantage.

Because the implementation of environmental regulations and management is still in a state of political infancy, it is more a question of how and to what extent the enlightened producer will alter important locational decisions with respect to present and perceived future environmental constraints. This policy domain has immense uncertainties; even the issue of what is long range is unclear. The World Association of Major Metropolises met in Japan in February 1994 at a special symposium with the title "Consideration of the Environment in the Urban Development: Towards the 21st Century." Their basic, yet complex, question can be summed up as follows: "How can the long-term development of major cities (especially during a period of economic change) be reconciled with respect for the urban environment, keeping in mind that these two concerns are closely linked?" (*Metropolis,* 1993).

One of the most promising set of environmental initiatives involves transnational cooperation, planning, and development between cities in different countries at different levels of development but with similar problems (water quality, traffic congestion, brownfields, etc.). The cities that are the most concerned about their comparative advantage in the 21st century will combine their strategic planning for economic development with their strategic planning for environmental management. Cities also need to recognize the issues of social development in a turbulent time of economic dislocations and readjustments.

Despite the uncertainties concerning the global implementation of environmental policies, two additional tentative propositions can be stated.

Proposition #9: International global environmental policies will develop slowly and uncertainly during the next two decades (1995 to 2015) and initially will be concerned with the macroproblems of oceans, global warming, rain forests, ozone layer, and so forth—the green agenda.

The magnitude of enacting and enforcing global green environmental policies is more and more apparent, but there is ready agreement about the nature and scope of the issues. The mechanisms for global implementation are still being developed, but that future, if not in sight, is at least known to be in the location "where dragons dwell."

The brown agenda is less well established. The gross problems of urban environments are of course well known, but the issues of environmental equity and the enhancement of the quality of urban life are only beginning to be articulated. The civic institutions of cities and their urban communities and cultures have evolved in different socioeconomic directions.

Urban-industrial society is experiencing a major paradigm shift. Although some cities are still chasing smokestacks, it is clear to others that "the commitment to maximum growth is being replaced by a desire to increase the quality of life and to preserve the environment. . . . One of the most important shifts is to commit to environmental balance rather than economic growth" (Theobald, 1992, p. 72).

Proposition #10: The cities that address their brown agenda earlier than others will enhance their long-term strategic comparative advantages.

With the expansion of the global environmental movement in the next decade, the challenge is to learn how to design the urban future so that the brown agenda is adequately addressed before too much damage occurs. Local decision making, local assessment, and local action planning will require a learning curve among municipal elites. The dynamic needs of economic productivity and environmental responsibility will require new forms of decision making by the civic elites throughout the urban agglomerations. That is the future, at least the optimistic version of an economically rational scenario, for a world order that respects the need to protect the environment. The empirical evidence is not yet established. But some portents of the future can be identified in relationship to Mexico and the environmental side agreements (Hopp, 1994).

Conclusions

The primary purpose of this book is to consider the changing role of cities as major actors on the stage of globalization. The concepts of intermestic policies and urban paradiplomacy are important conceptual constructs to the future of cities and their development policies as they approach the 21st century and the challenges and opportunities of the decade ahead. The authors in this volume evaluate the prospects of North American cities with ample empirical information and theoretical analysis. Their analyses and information provide an introduction to the cities seeking to claim a major role in the evolving and expanding global economy. The importance of functional cooperation, interaction, and networking are illustrated for a number of cities and regions. The dark side of free trade for cities is also introduced with a concern for economic dislocation and inequality and the degradation of the environment. If liberalized trade enhances consumer sovereignty, concern must also be expressed for what happens to the consumer in his or her role as citizen and producer.

The 10 propositions developed in this epilogue are proposed as a way to monitor the urban consequences of liberalized trade in that portion of the global society that has become substantially economically interconnected. In the next two decades, the economic competition between cities in that sphere will only become more intense. In the long term, the real winners will be those communities that are also able to make significant strategic investment and commitments to their global competitiveness, with appropriate respect for the quality of life for all their citizens, for the upgrading of their infrastructure, and for the protection and enhancement of their environment and its ecosystems.

NOTES

1. For a good summary of recent Western economic history, see Palmer and Cotton (1992).

2. The brown agenda consists of concerns with brown air, brown clouds, brown water, brown toxic sites (*brownfields*), and multicolored urban solid waste, all resulting from urbanization and its physical and economic embodiments in specific cities (Gappert, 1993).

REFERENCES

Ahmal, J. (1978). *Import substitution, trade and development.* Greenwich, CT: JAI.

Balassa, B. (1981). *The process of industrial development and alternative development strategies.* Princeton, NJ: Princeton University, Department of Economics, International Finance Section.

Brown, L. (1993). *State of the world 1993.* Washington, DC: World Watch Institute.

Chenery, H. (1979). *Structural change and development policy.* London: Oxford University Press.

Dorfman, R., & Dorfman, N. S. (1972). *Economics of the environment.* New York: Norton.

Drucker, P. (1994, November). The age of social transformation. *Atlantic Monthly, 274,* 53-80.

Gappert, G. (1993, November). The future of urban environments: Implications for the business community. *Business Horizons, 37*(6).

Harbrect, D. (1994, November 21). What has NAFTA wrought? Plenty of trade. *Business Week,* 48-49.

Hirschman, A. O. (1958). *The strategy of economic development.* New Haven, CT: Yale University Press.

Hopp, R. (1994). The new era of global environmental protection: NAFTA "1, 2, 3." *Journal of Environmental Regulation, 3,* 233-250.

Jacobs, J. (1976). *The economy of cities.* New York: Random House.

Kindleberger, C. P. (1964). *Foreign trade and the national economy.* New Haven, CT: Yale University Press.

Knight, R. V. (1989). City development and urbanization: Building the knowledge-based city. In R. V. Knight & G. Gappert (Eds.), *Cities in a global society* (Urban Affairs Annual Review, Vol. 35, pp. 223-242). Newbury Park, CA: Sage.

Knight, R. V. (1992).*The future of European cities: The role of science and technology.* Brussels, Belgium: Commission of the European Communities.

Knight, R. V., & Gappert, G. (Eds.). (1989). *Cities in a global society* (Urban Affairs Annual Review, Vol. 35). Newbury Park, CA: Sage.

Lemberg, K. (1989). The need for autonomy: Improving local democracy. In R. V. Knight & G. Gappert (Eds.), *Cities in a global society* (Urban Affairs Annual Review, Vol. 35, pp. 207-222). Newbury Park, CA: Sage.

Maizels, A. (1963). *Industrial growth and world trade.* London: Cambridge University Press.

Meadows, D., et al. (1972). *The limits to growth: A report for the Club of Rome.* New York: Universe.

Metropolis. (1993, November). [Newsletter]. International Association of Major Metropoles, Paris, France.

Paauw, D. (1970). *Development strategies in open dualistic economies.* Washington, DC: National Planning Association.

Paauw, D., & Fei, J. (1973). *The transition in open dualistic economies.* New Haven, CT: Yale University Press.

Palmer, R. R., & Cotton, J. (1992). Recession and recovery: The global economy. In R. R. Palmer & J. Cotton (Eds.), *A history of the modern world* (pp. 982-1019). New York: McGraw-Hill.

Schumpeter, J. A. (1967). *Economic doctrine and method: An historical sketch.* New York: Oxford University Press.

Soldatos, P. (1991) Strategic cities alliances: An added value to the innovative making of an international city. *Ekistics, 58*(350-351), 346-350.

Soldatos, P. (1993, October). *Strategic cities network in the European Community.* Paper presented at the New International Cities Era (NICE) conference, David Kennedy Center for International Studies, Brigham Young University, Provo, UT.

Theobald, R. (1992). *Turning the century.* Indianapolis, IN: Knowledge Press.

Thompson, W. (1968). *A preface to urban economics.* Baltimore: Johns Hopkins University Press.

World Commission on Environment and Development. (1976). *Our common future.* New York: Oxford University Press.

Zuvekas, C. (1979). *Economic development: An introduction.* New York: St. Martin's.

A Select Bibliography on Cities and Globalization

PAUL GREGORY

As the urban environment faces heightened pressures from economic restructuring, increasingly sophisticated telecommunications, movement of capital, and the mass migration of labor, a recognition exists that macroeconomic policies and international political circumstances animate local-level decisions. During the past 10 years, the evolving body of literature dealing with the city and globalization represents an amalgamation of two distinct and disparate literatures.

Until recently, the broader globalization literature has drawn primarily from research on international economic activity and capital flows focusing on key corporate players—banks and multinationals. The key element in this perspective is the multinational corporation as the principal player in the process of globalization.

City issues have, until recently, largely been ignored in globalization research. The literature on cities and urban areas has traditionally focused internally and has been interpreted in terms of national and regional political, economic, and social systems.

The relatively recent realization of the city as a key stakeholder in the world economic order has focused attention more specifically on the importance of local political and economic conditions in the process of globalization. The primary producers of this multidisciplinary literature are researchers who have been described as the "World Cities Group." Writing from a predominantly North American and Western European perspective, the world cities literature focuses primarily on qualitative and sociological research. Telecommunications (Castells, 1992; Moss, 1987b); economic restructuring (Kasaba, 1991; Sassen, 1994a, 1994b, 1994c; M. P. Smith, 1987); the international flow of financial capital (Fagan, 1994; N. Fainstein, S. Fainstein, & A. Schwartz, 1989; Feagin & Smith, 1987; Sassen, 1994a;

Thrift, 1987); and the mass migration of labor (Sassen, 1988) are all issues considered to be important. Some of these studies are comparative (Dogan & Kasarda, 1988a, 1988b; P. Hall, 1993; Sassen, 1990b); others isolate specific urban centers (Daly & Stimson, 1991; K. Fujita, 1991). The majority of the research, however, addresses cities and globalization issues in a more general way (Chase-Dunn, 1989; Freidmann, 1993; Knight & Gappert, 1989; Logan & Swanstrom, 1990a, 1990b; Mollenkopf, 1993b; Timberlake, 1987).

Largely lacking from the literature are transnational studies—studies that focus on the development of cities, urban systems, and their interconnectedness. Also conspicuously absent from the literature is any substantial contribution by researchers from, and working in, emerging nations. Research that is done on the developing world tends to focus not on globalization as such but on conditions and urban attributes of major cities.

As pressures continue to mount on cities, research on the interdependence of globalization and city development will continue to expand and evolve. The past 10 years of research have been characterized by considerable fragmentation, and the literature of globalization remains as broad as the definition of the concept itself. In the years ahead, researchers look forward to increasingly systematic and evaluative studies of this complex set of issues and, specifically, of the role of individual cities and systems of cities in the globalization process.

Alger, C. F. (1990). The world relations of cities. *International Studies Quarterly, 34,* 493-518.

Amin, A., & Thrift, N. (1992). Neo-Marshallian nodes in global networks. *International Journal of Urban and Regional Research, 16,* 571-587.

Beauregard, R. A. (1989a). *Atop the urban hierarchy.* Totowa, NJ: Rowman & Littlefield.

Beauregard, R. A. (1989b). *Economic restructuring and political response.* Newbury Park, CA: Sage.

Bestor, T. (1989). *Neighborhood Tokyo.* Stanford, CA: Stanford University Press.

Bingham, R. D., & Blair, J. P. (Eds.). (1984). *Urban economic development* (Urban Affairs Annual Review, Vol. 27). Beverly Hills, CA: Sage.

Black, T. J. (1981). The changing economic role of central cities and suburbs. In A. Solomon (Ed.), *The prospective city* (pp. 80-123). Cambridge: MIT Press.

Blakely, E. J. (1994). *Transformation of the American metropolis* (Working Paper 615). Berkeley: University of California, Institute of Urban and Regional Development.

Blakely, E. J., & Stimson, R. J. (1992a). Interdependencies and the new urban form in the Pacific Rim cities. In E. J. Blakely & R. J. Stimson (Eds.), *New cities of the Pacific Rim* (Monograph 43, pp. 1.1-1.32). Berkeley: University of California, Institute of Urban and Regional Development.

Blakely, E. J., & R. J. Stimson (Eds.). (1992b). *New cities of the Pacific Rim* (Monograph 43). Berkeley: University of California, Institute of Urban and Regional Development.

Boulding, K. (1978). The city as an element in the international system. In L. S. Bourne & J. W. Simmons (Eds.), *Systems of cities* (pp. 150-168). New York: Oxford University Press.

Bradshaw, Y. W. (1987). Urbanization and underdevelopment: A global study of urbanization, global bias and economic dependency. *American Sociological Review, 52,* 224-239.

Brecher, C., & Horton, R. D. (Eds.). (1987). *Setting municipal priorities 1988.* New York: New York University Press.

Bromley, R., & Birkbeck, C. (1988). Urban economy and employment. In M. Pacione (Ed.), *The geography of the third world* (pp. 114-147). London: Routledge.

Bruyn, S. T., & Meehan, J. (Eds.). (1987). *Beyond the market and the state: New directions in community development.* Philadelphia: Temple University Press.

Buck, N., Drennan, M., & Newton, K. (1992). Dynamics of the metropolitan economy. In S. Fainstein, I. Gordon, & M. Harloe (Eds.), *Divided cities: New York and London in the contemporary world* (pp. 68-104). London: Basil Blackwell.

Budd, L., & Whimster, S. (Eds.). (1992). *Global finance and urban living: A study of metropolitan change.* London: Routledge.

Burgel, G. (1993). *La ville aujourd'hui* [The present-day city] (Collection Pluriel). Paris: Hachette.

Castells, M. (1987). Technoeconomic restructuring, socio-political processes and spatial transformation: A global perspective. In J. Henderson & M. Castells (Eds.), *Global restructuring and territorial development* (pp. 1-17). Newbury Park, CA: Sage.

Castells, M. (1989). *The informational city: Information technology, economic restructuring, and the urban regional process.* London: Basil Blackwell.

Castells, M. (1992). *European cities, the informational society and the global economy.* Amsterdam: Amsterdam University, Center for Metropolitan Research.

Chandler, J. A. (1991). *Local government today.* Manchester, UK: Manchester University Press.

Changrien, P., & Stimson, R. J. (1992). Bangkok: Jewel in Thailand's crown. In E. J. Blakely & R. J. Stimson (Eds.), *New cities of the Pacific Rim* (Monograph 43, pp. 15.1-15.39). Berkeley: University of California, Institute of Urban and Regional Development.

Chase-Dunn, C. (1984). Urbanization in the world system: New directions for research. In M. P. Smith (Ed.), *Cities in transformation* (pp. 111-120). Beverly Hills, CA: Sage.

Chase-Dunn, C. (1985). The system of world cities. In M. Timberlake (Ed.), *Urbanization in the world economy* (pp. 269-292). New York: Academic Press.

Chase-Dunn, C. (1989). *Global formation.* Cambridge, MA: Blackwell.

Chase-Dunn, C., & Willard, A. (1993, April). *Cities in the central world system since* AD *1200: Size, hierarchy and domination.* Paper presented at the conference on World Cities in a World System, Sterling, VA.

Church, G. (1994). The new economy, the new cities: Prospect. In E. Isin (Ed.), *Toronto: Region in the world economy: A symposium on how the Toronto region is affected by changes in the world economy* (pp. 223-229). North York, ON: York University, Urban Studies Programme.

Cisneros, H. G. (Ed.). (1993). *Interwoven destinies: Cities and the nation.* New York: Norton.

Clark, T. N., & Inglehart, R. (1990). *The new political culture.* Paper presented at the biennial meeting of the International Sociological Association, Madrid, Spain.

Clarke, S. E. (Ed.). (1989). *Urban innovation and autonomy: Political implications of policy change.* Newbury Park, CA: Sage.

Clarke, S. E., & Kirby, A. (1990). In search of the corpse: The mysterious case of local politics. *Urban Affairs Quarterly, 25,* 389-412.

Clavel, P. (1986). *The progressive city.* New Brunswick, NJ: Rutgers University Press.

Cochrane, A., & Clarke, J. (Eds.). (1993). *Comparing welfare states: Britain in an international context.* London: Sage.

Cohen, R. B. (1981). The new international division of labor multinational corporations and urban hierarchy. In M. Dear & A. Scott (Eds.), *Urbanization and urban planning in capitalist society* (pp. 287-318). New York: Methuen.

Corey, K. E. (1987). Planning the information age metropolis: The case of Singapore. In L. Guelke & R. Preston (Eds.), *Abstract thoughts: Concrete solutions* (Publications Series, No. 29, pp. 49-72). Waterloo, ON: University of Waterloo, Department of Geography.

Corey, K. E., Fletcher, R. G., & Moscove, B. J. (1992). Singapore: The planned new city of the Pacific Rim. In E. J. Blakely & R. J. Stimson (Eds.), *New cities of the Pacific Rim* (Monograph 43, pp. 14.1-14.31). Berkeley: University of California, Institute of Urban and Regional Development.

Cox, K. R. (1991a). The abstract, the concrete and argument in the new urban politics. *Journal of Urban Affairs, 13,* 299-306.

Cox, K. R. (1991b). Questions of abstraction in studies in the new urban politics. *Journal of Urban Affairs, 13,* 267-280.

Cox, K. R. (1992). The politics of globalization: A sceptic's view. *Political Geography, 11,* 427-429.

Cox, K. R. (1993). The local and the global in the new urban politics: A critical view. *Environment and Planning D: Society and Space, 11,* 433-448.

Cox, K. R., & Mair, A. J. (1988). Locality and community in the politics of local economic development. *Annals of the Association of American Geographers, 78,* 307-325.

Cox, K. R., & Mair, A. J. (1989). Urban growth machines and the politics of local economic development. *International Journal of Urban and Regional Research, 13*(1), 137-146.

Cox, K. R., & Mair, A. J. (1991). From localised social structures to localities as agents. *Environment and Planning A, 23,* 197-213.

Daly, M. T., & Stimson, R. J. (1991). *The internationalisation of Australian cities.* Brisbane: Australian Research Council Grant.

Daly, M. T., & Stimson, R. J. (1992). Sydney: Australia's gateway and financial capital. In E. J. Blakely & R. J. Stimson (Eds.), *New cities of the Pacific Rim* (Monograph 43, pp. 18.1-18.42). Berkeley: University of California, Institute of Urban and Regional Development.

Daniels, P. L .D. (1991). *Services and metropolitan development: International perspectives.* London: Routledge.

Daniels, P. W. (1990). *Change and transition in metropolitan areas: The role of tertiary industries working group, world association of metropolises.* Plymouth, UK: Service Industries Research Centre.

Dear, M., & Scott, A. (Eds.). (1981). *Urbanization and urban planning in capitalist society.* New York: Methuen.

Dicken, P. (1992). *Global shift: The internationalization of economic activity.* New York: Guilford.

DiGaetano, A. (1989). Urban political regime formation: A study and contrast. *Journal of Urban Affairs, 11,* 261-282.

Dinford, M., & Kafkala, G. (Eds.). (1992). *Cities and regions in the New Europe: The global-local interplay and spatial development strategies.* London: Bellhaven.

Doeringer, P. B., Terkla, D. G., & Topakian, G. C. (1988). *Invisible factors in local economic development.* New York: Oxford University Press.

Dogan, M., & Kasarda, J. (1988a). *The metropolis era.* Newbury Park, CA: Sage.

Dogan, M., & Kasarda, J. (Eds.). (1988b). *A world of giant cities* (The Metropolitan Era, Vol. 1). Newbury Park, CA: Sage.

Douglas, M. (1987a). *The future of cities on the Pacific Rim.* Honolulu: University of Hawaii, Department of Urban and Regional Planning.

Douglas, M. (1989). The future of cities on the Pacific Rim. In M. P. Smith (Ed.), *Pacific Rim cities in the world economy* (Comparative and Community Research Series, Vol. 2, pp. 9-67). New Brunswick, NJ: Transaction.

Drache, D. (1994). Globalization: The dilemma of nations. In E. Isin (Ed.), *Toronto: Region in the world economy: A symposium on how the Toronto region is affected by changes in the world economy* (pp. 16-21). North York, ON: York University, Urban Studies Programme.

Drache, D., & Gertler, M. (Eds.). (1991). *The new era of global competition: State policy and market power.* Montreal, PQ: McGill-Queen's University Press.

Drakis-Smith, D. (Ed.). (1986). *Urbanisation in the developing world.* London: Croom Helm.

Drennan, M. P. (1992a). Gateway cities: The metropolitan sources of U.S. producer service exports. *Urban Studies, 29*(2), 217-235.

Drennan, M. P. (1992b). New York in the world economy. *Survey of Regional Literature, 7,* 7-12.

Drucker, P. (1991). The changed world economy. In S. Fosler (Ed.), *Local economic development.* Washington, DC: International City Management Association.

Duncan, S. S., & Goodwin, M. (1988). *The local state and uneven development.* Oxford, UK: Blackwell.

Dunning, J. H., & Norman, G. (1987). The location choice of offices of international companies. *Environment and Planning A, 19,* 613-631.

Eaton, L. (1989). *Gateway cities and other essays.* Ames: Iowa State Press.

Edel, M. (1981). Capitalism, accumulation and the explanation of urban phenomena. In M. Dear & A. Scott (Eds.), *Urbanization and urban planning in capitalist society* (pp. 19-44). New York: Methuen.

Edgington, D. W., & Goldberg, M. A. (1992). Vancouver: Canada's gateway to the rim. In E. J. Blakely & R. J. Stimson (Eds.), *New cities of the Pacific Rim* (Monograph 43, pp. 7.1-7.29). Berkeley: University of California, Institute of Urban and Regional Development.

Elkin, S. L. (1985). Twentieth century urban regimes. *Journal of Urban Affairs, 7,* 11-28.

Elkin, S. L. (1987). *City and regime in the American Republic.* Chicago: University of Chicago Press.

European Institute of Urban Affairs. (1992). *Urbanisation and the functions of cities in the European community: A report to the Commission of European Communities, Directorate General for Policy (XVI).* Liverpool, UK: John Moores University.

Ewers, H.-J., Goddard, J. B., & Matzerath, H. (Eds.). (1986). *The future of the metropolis: Berlin, Paris, London, New York.* Berlin: Walter de Gruyter.

Fagan, R. H. (1994). Reinterpreting the geography of capital. *Environment and Planning D: Society and Space, 12,* 265-285.

Fainstein, N. I., & Fainstein, S. S. (1982). *Urban policy under capitalism* (Urban Affairs Annual Review, Vol. 22). Beverly Hills, CA: Sage.

Fainstein, N. I., Fainstein, S. S., & Schwartz, A. (1989). Economic shifts and land use in the global city: New York, 1940-1987. In R. A. Beauregard (Ed.), *Atop the urban hierarchy* (pp. 45-85). Totowa, NJ: Rowman & Littlefield.

Fainstein, S. S. (1991). Rejoinder to: Questions of abstraction in studies in the new urban politics. *Journal of Urban Affairs, 13,* 281-288.

Fainstein, S. S. (1993). *The city builders.* Oxford, UK: Blackwell.

Fainstein, S. S., Fainstein, N. I., Hill, R. C., Judd, D. R., & Smith, M. P. (1983). *Restructuring the city.* New York: Longman.

Fainstein, S. S., Gordon, I., & Harloe, M. (1992a). *Divided cities: Economic restructuring and social change in London and New York.* London: Basil Blackwell.

Fainstein, S. S., Gordon, I., & Harloe, M. (Eds.). (1992b. *Divided cities: New York and London in the contemporary world.* London: Basil Blackwell.

Feagin, J. R. (1987). The secondary circuit of capital. *International Journal of Urban and Regional Research, 11*(2), 171-192.

Feagin, J. R. (1988). *Free enterprise city.* New Brunswick, NJ: Rutgers University Press.

Feagin, J. R., & Smith, P. S. (Eds.). (1987). *The capitalist city: Global restructuring and community politics.* Oxford, UK: Basil Blackwell.

Fisher, R., & Kling, J. (Eds.). (1993). *Mobilizing the community: Local politics in the era of the global city.* (Urban Affairs Annual Review, Vol. 41). Newbury Park, CA: Sage.

Forbes, D. (1992). The internationalization of Jakarta and the growth of the services sector. In E. J. Blakely & R. J. Stimson (Eds.), *New cities of the Pacific Rim* (Monograph 43, pp. 16.1-16.22). Berkeley: University of California, Institute of Urban and Regional Development.

Freidmann, J. (1986). The world city hypothesis. *Development and Change, 17*(1), 69-83.

Freidmann, J. (1993, April). *Where we stand: A decade of world city research.* Paper presented at the conference on World Cities in a World System, Sterling, VA.

Freidmann, J., & Wolff, G. (1982). World city formation: An agenda for research and action. *International Journal of Urban and Regional Research, 6*(3), 309-344.

Frost, M., & Spence, N. (1992). Global city characteristics and central London's employment. *Urban Studies, 30*(3), 547-558.

Fujita, K. (1988). The technopolis: High technology and regional development in Japan. *International Journal of Urban and Regional Research, 12*(4), 566-594.

Fujita, K. (1991). A world city and flexible specialisation: Restructuring of the Tokyo metropolis. *International Journal of Urban and Regional Research, 15*(1), 269-284.

Fujita, M. (1993). Monopolistic competition and urban systems. *European Economic Review, 37*(2-3), 308-315.

Fuller, S. S. (1989). The internationalization of the Washington, D.C., area economy. In R. V. Knight & G. Gappert (Eds.), *Cities in a global society* (Urban Affairs Annual Review, Vol. 35, pp. 108-119). Newbury Park, CA: Sage.

Ganz, A., & Konga, L. F. (1989). Boston in the world economy. In R. V. Knight & G. Gappert (Eds.), *Cities in a global society* (Urban Affairs Annual Review, Vol. 35, pp. 132-140). Newbury Park, CA: Sage.

Gappert, G. (1989a). Global thinking and urban planning. In R. V. Knight & G. Gappert (Eds.), *Cities in a global society* (Urban Affairs Annual Review, Vol. 35, pp. 305-311). Newbury Park, CA: Sage.

Gappert, G. (1989b). A management perspective on cities in a changing global environment. In R. V. Knight & G. Gappert (Eds.), *Cities in a global society* (Urban Affairs Annual Review, Vol. 35, pp. 312-325). Newbury Park, CA: Sage.

Garau, P. (1989). Third world cities in a global society. In R. V. Knight & G. Gappert (Eds.), *Cities in a global society* (Urban Affairs Annual Review, Vol. 35, pp. 68-78). Newbury Park, CA: Sage.

Garreau, J. (1991). *Edge city.* New York: Doubleday.

Gertler, L. (1989). Telecommunications and the changing global context of urban settlements. In R. V. Knight & G. Gappert (Eds.), *Cities in a global society* (Urban Affairs Annual Review, Vol. 35, pp. 272-284). Newbury Park, CA: Sage.

Gibson, D. V., Kozmetsky, G., & Smilor, R. (1992). *The technopolis phenomenon: Smart cities, fast systems and global networks.* Baltimore: Rowman & Littlefield.

Gilb, C. L. (1989). Third world cities: Their role in the global economy. In R. V. Knight & G. Gappert (Eds.), *Cities in a global society* (Urban Affairs Annual Review, Vol. 35, pp. 96-107). Newbury Park, CA: Sage.

Glazer, N. (1994). *New York: The fate of a world city?* Washington, DC: Woodrow Wilson International Center for Scholars.

Goddard, J. B. (1993). *Information and communications technologies, corporate hierarchies and urban hierarchies in the new Europe.* Paper presented at the Fourth International Workshop on Technological Change and Urban Form: Productive and Sustainable Cities, Berkeley, CA.

Goetz, E. G., & Clarke, S. E. (Eds.). (1993). *The new localism: Comparative urban politics in a global era* (Sage Focus Editions, Vol. 164). Newbury Park, CA: Sage.

Goldberg, M. A., Helsley, R., & Levi, M. (1988). *Factors influencing the development of financial centers.* Vancouver: University of British Columbia, Faculty of Commerce and Business Administration.

Goldberg, M. A., Helsley, R., & Levi, M. (1989). The location of international financial activity. *Regional Studies, 23,* 1-10.

Gordon, D. M. (1988). The global economy: New edifice or crumbling foundations. *New Left Review, 168,* 24-64.

Gottdiener, M. (Ed.). (1986). *Cities in stress: A new look at the urban crisis* (Urban Affairs Annual Review, Vol. 30). Beverly Hills, CA: Sage.

Gottdiener, M. (1987). *The decline of urban politics.* Newbury Park, CA: Sage.

Gottman, J. (1983). *The coming of the transactional city.* College Park: University of Maryland, Institute for Urban Studies.

Gottman, J. (1989). What are cities becoming centers of? Sorting out the possibilities. In R. V. Knight & G. Gappert (Eds.), *Cities in a global society* (Urban Affairs Annual Review, Vol. 35, pp. 58-67). Newbury Park, CA: Sage.

Grosfoguel, R. (1993, April). *World cities in the Caribbean city system: Miami and San Juan.* Paper presented at the conference on World Cities in a World System, Blacksburg, VA.

Gurr, T. R., & King, D. (1987). *The state and the city.* Chicago: University of Chicago Press.

Hall, P. (1966). *World cities.* New York: McGraw-Hill.

Hall, P. (1988). *Cities of tomorrow.* Oxford, UK: Blackwell.

Hall, P. (1993). Forces shaping urban Europe. *Urban Studies, 30*(6), 883-898.

Hall, P., & Hay, D. (1980). *Growth centres in the European urban system.* Berkeley: University of California Press.

Hall, S. (1991). The local and the global: Globalization and ethnicity. In A. D. King (Ed.), *Culture, globalization and the world-system: Contemporary conditions for the representation of identity* (Current Debates in Art History, Vol. 3). Binghampton: State University of New York, Department of Art and Art History.

Harvey, D. (1985). *The urbanisation of capital.* Oxford, UK: Basil Blackwell.

Henderson, J., & Castells, M. (Eds.). (1987). *Global restructuring and territorial development.* Newbury Park, CA: Sage.

Hepworth, M. E. (1990). *Geography of the information economy* (Center for Urban and Regional Studies Series). New York: Guilford.

Herzog, L. (1990). *Where north meets south: Cities, space and politics on the U.S.-Mexico border.* Austin, TX: CMAS.

Herzog, L. (1991). Cross-national urban structure in the era of global cities: The U.S. and Mexico. *Urban Studies, 28*(4), 519-533.

Herzog, L. (1992). The U.S.-Mexico transfrontier metropolis. *Business Mexico, 2*(4), 38-43.

Hicks, D. A. (1983). Urban and economic adjustment to the post-industrial era. In D. A. Hicks & N. J. Glickman (Eds.), *Transition to the 21st century: Prospects and policies for economic and urban-regional transformation* (pp. 345-370). Greenwich, CT: JAI.

Hicks, D. A., & Glickman, N. J. (Eds.). (1983). *Transition to the 21st century: Prospects and policies for economic and urban-regional transformation.* Greenwich, CT: JAI.

Hill, R. (1984). Urban political economy: Emergence, consolidation and development. In M. P. Smith (Ed.), *Cities in transformation* (pp. 123-137). Beverly Hills, CA: Sage.

HMSO. (1991). *London: World city moving into the 21st century.* London: Author.

Hollis, G., Ham, G., & Ambler, M. (Eds.). (1992). *The future role and structure of local government.* Harlow, UK: Longman.

Hong, S. W. (1992). Seoul dominates Korea's emergence. In E. J. Blakely & R. J. Stimson (Eds.), *New cities of the Pacific Rim* (Monograph 43, pp. 11.1-11.16). Berkeley: University of California, Institute of Urban and Regional Development.

Howard, J. (1989). Long wave cycles and cities in a global economy. In R. V. Knight & G. Gappert (Eds.), *Cities in a global society* (Urban Affairs Annual Review, Vol. 35, pp. 169-180). Newbury Park, CA: Sage.

Howell, H. (1988). *Economic, technological and locational trends in European cities.* Aldershot, UK: Commission on the European Communities, Fast Program.

Isin, E. (1994a). The dangers of discourse on the global city. In E. Isin (Ed.), *Toronto: Region in the world economy: A symposium on how the Toronto region is affected by changes in the world economy* (pp. 205-210). North York, ON: York University, Urban Studies Programme.

Isin, E. (Ed.). (1994b). *Toronto: Region in the world economy: A symposium on how the Toronto region is affected by changes in the world economy.* North York, ON: York University, Urban Studies Programme.

Ito, T., & Tanifuji, M. (1982). The role of small and intermediate cities in national development in Japan. In O. P. Mathur (Ed.), *Small cities and national development* (pp. 22-35). Nagoya, Japan: UNCRD.

Jacobs, B. D. (1992). *Fractured cities: Capitalism, community and empowerment in Britain and America.* London: Routledge.

Johnston, R. J. (1985). Local government and the state. In M. Pacione (Ed.), *Progress in political geography* (pp. 152-176). London: Croom Helm.

Judd, D., & Parkinson, M. (Eds.). (1990). *Leadership and urban regeneration: Cities in North America and Europe.* Newbury Park, CA: Sage.

Judd, D., Parkinson, M., & Foley, B. (Eds.). (1989). *Regenerating the cities.* New York: Scott, Foresman.

Jussawalla, M., & Cheah, C. W. (1983). Towards an information economy: The case of Singapore. *Information Economics and Policy, 1,* 161-176.

Kantor, P., & Stephen, D. (1988). *The dependent city: The changing political economy of Urban America.* Chicago: Scott, Foresman.

Kasaba, R. (Ed.). (1991). *Cities in the world-system* (Studies in the Political Economy of the World-System, Vol. 126). New York: Greenwood.

Kasarda, J. D., & Crenshaw, E. M. (1991). Third world urbanization: Dimensions, theories and determinants. *Annual Review of Sociology, 17,* 467-501.

Keil, R., & Ronneberger, K. (1994). Going up the country: Internationalisation and urbanisation on Frankfurt's northern fringe. *Environment and Planning D: Society and Space, 12,* 58-73.

Kincaid, A. D., & Portes, A. (Eds.). (1994). *Comparative national development: Sociological perspectives for the new global order.* Chapel Hill: University of North Carolina Press.

King, A. D. (1984). *The bungalow: The production of a global culture.* London: Routledge & Kegan Paul.

King, A. D. (1985). Colonial cities: Global pivots of change. In R. Ross & G. Telkamp (Eds.), *Colonial cities* (pp. 7-32). Dordrecht, The Netherlands: Martinus Nijhoff for the Leiden University Press.

King, A. D. (1989). Colonialism, urbanism, and the capitalist world economy: An introduction. *International Journal of Urban and Regional Research, 13*(1), 1-18.

King, A. D. (1990). *Global cities: Post imperialism and the internationalisation of London.* London: Routledge.

King, A. D. (Ed.). (1991a). *Culture, globalization and the world system.* Binghamton: State University of New York, Department of Art and History.

King, A. D. (Ed.). (1991b). *Culture, globalization and the world-system: Contemporary conditions for the representation of identity* (Current Debates in Art History, Vol. 3). Binghamton: State University of New York, Department of Art and Art History.

King, D. S., & Pierre, J. (Eds.). (1990). *Challenges to local government.* London: Sage.

Kirby, A. (1985). Nine fallacies of local economic change. *Urban Affairs Quarterly, 21,* 207-220.

Kirby, A. (1993). *Power/resistance: Local politics and the chaotic state.* Bloomington: Indiana University Press.

Klaassen, L., Van der Berg, L., & Van Der Meer, J. (1989). *The city: Engine behind the economic recovery.* Brookfield, MA: Avebury.

Knight, R. V., & Gappert, G. (1984). Cities and the challenge of the global economy. In R. D. Bingham & J. P. Blair (Eds.), *Urban economic development* (Urban Affairs Annual Review, Vol. 27, pp. 63-78). Beverly Hills, CA: Sage.

Knight, R. V. (1986). The advanced industrial metropolis: A new type of world city. In H.-J. Ewers, J. B. Goddard, & H. Matzerath (Eds.), *The future of the metropolis: Berlin, Paris, London, New York* (pp. 391-436). Berlin: Walter de Gruyter.

Knight, R. V. (1989a). City building in global society. In R. V. Knight & G. Gappert (Eds.), *Cities in a global society* (Urban Affairs Annual Review, Vol. 35, pp. 326-334). Newbury Park, CA: Sage.

Knight, R. V. (1989b). The emergent global society. In R. V. Knight & G. Gappert (Eds.), *Cities in a global society* (Urban Affairs Annual Review, Vol. 35, pp. 24-43). Newbury Park, CA: Sage.

Knight, R. V., & Gappert, G. (Eds.). (1989). *Cities in a global society* (Urban Affairs Annual Review, Vol. 35). Newbury Park, CA: Sage.

Kosambi, M. (1985). Commerce, conquest and the colonial city: The role of locational factors in the rise of Bombay. *Economic and Political Weekly, 5,* 31-37.

Lampard, E. E. (1986). The New York metropolis in transformation: History and prospect. In H.-J. Ewers, J. B. Goddard, & H. Matzerath (Eds.), *The future of the metropolis: Berlin, Paris, London, New York* (pp. 27-110). Berlin: Walter de Gruyter.

Latouche, D. (1994). Localisation: The dilemma of cities. In E. Isin (Ed.), *Toronto: Region in the world economy: A symposium on how the Toronto region is affected by changes in the world economy* (pp. 22-24). North York, ON: York University, Urban Studies Programme.

Lawson, V., & Klak, T. (1990). Conceptual linkages in the study of production and reproduction in Latin American cities. *Economic Geography, 66*(4), 310-327.

Lee, R., & Pelizzon, S. (1991). Hegemonic cities in the modern world-system. In R. Kasaba (Ed.), *Cities in the world-system* (Studies in the Political Economy of the World-System, Vol. 126, pp. 43-54). New York: Greenwood.

Le Galès, P. (1990). Economic regeneration in Rennes: Local social dynamics and state support. In D. Judd & M. Parkinson (Eds.), *Leadership and urban regeneration: Cities in North America and Europe* (pp. 69-85). Newbury Park, CA: Sage.

Léviellée, J., & Whelan, R. K. (1990). Montreal: The struggle to become a world city. In D. Judd & M. Parkinson (Eds.), *Leadership and urban regeneration: Cities in North America and Europe* (pp.152-170). Newbury Park, CA: Sage.

Levine, M. V. (1989). Urban redevelopment in the global economy: The cases of Montreal and Baltimore. In R. V. Knight & G. Gappert (Eds.), *Cities in a global society* (Urban Affairs Annual Review, Vol. 35, pp. 141-152). Newbury Park, CA: Sage.

Light, I. (1983). *Cities in world perspective.* New York: Macmillan.

Linn, J. F. (1983). *Cities in the developing world: Policies for their equitable and efficient growth.* New York: Oxford University Press.

Logan, J. R., & Molotch, H. (1987). *Urban fortunes: The political economy of place.* Berkeley: University of California Press.

Logan, J. R., & Swanstrom, T. (1990a). Urban restructuring: A critical view. In J. R. Logan & T. Swanstrom (Eds.), *Beyond the city limits: Urban policy and the economic restructuring in comparative perspective* (pp. 67-90). Philadelphia: Temple University Press.

Logan, J. R., & Swanstrom, T. (Eds.). (1990b). *Beyond the city limits: Urban policy and the economic restructuring in comparative perspective.* Philadelphia: Temple University Press.

Logan, M. J., & Salih, K. K. (1982, January). *Implications of international economic adjustments for national urban development and policies.* Paper presented at the conference on Urbanization and National Development, East-West Center, Honolulu, HI.

Lyman, B. (1992). Urban primacy and world-system position. *Urban Affairs Quarterly, 28,* 22-37.

MacEwan, A. (1994). Globalization and stagnation. In R. Milibrand & L. Panitch (Eds.), *Between globalism and nationalism.* New York: Monthly Review Press.

Machimura, T. (1992). The urban restructuring process in the 1980s: Transforming Tokyo into a world city. *International Journal of Urban and Regional Research, 16*(1), 114-128.

Magdoff, H. (1993). *Globalization: To what end?* New York: Monthly Review Press.

Mandelbaum, S. J. (1986). Cities and telecommunications: The limits of community. *Telecommunications Policy, 10*(2), 58-77.

Manzo, K. A. (1992). *Domination, resistance, and social change in South Africa: The local effects of global power.* New York: Praeger.

Markusen, A. (1987). *Regions: The economics and politics of territory.* Totowa, NJ: Rowman & Littlefield.

Marlin, J. T., Ness, I., & Collins, S. T. (1986). *Book of world city rankings.* New York: Macmillan.

Masai, Y. (1989). Greater Tokyo as a global city. In R. V. Knight & G. Gappert (Eds.), *Cities in a global society* (Urban Affairs Annual Review, Vol. 35, pp. 153-164). Newbury Park, CA: Sage.

Mathur, O. P. (Ed.). (1982). *Small cities and national development.* Nagoya, Japan: UNCRD.

McGee, T. G. (1984, January). *Circuits and networks of capital: The internationalisation and the world economy and national urbanisation.* Paper presented at the conference on Urban Growth and Economic Development in the Pacific Region, Institute of Economics, Academia Sinica, Taipei, Taiwan.

Mera, K. (1992). Tokyo: A world capital in Asia. In E. J. Blakely & R. J. Stimson (Eds.), *New cities of the Pacific Rim* (Monograph 43, pp. 9.1-9.18). Berkeley: University of California, Institute of Urban and Regional Development.

Meyer, D. R. (1991a). Change in the world system of metropolises: The role of business intermediaries. *Urban Geography, 12*(5), 393-417.

Meyer, D. R. (1991b). The formation of a global financial center: London and its intermediaries. In R. Kasaba (Ed.), *Cities in the world-system* (Studies in the Political Economy of the World-System, Vol. 126, pp. 97-106). New York: Greenwood.

Mingione, E. (1991). *Fragmented societies: A sociology of economic life beyond the market paradigm.* Oxford, UK: Blackwell.

Miyakawa, Y. (1983). Metamorphosis of the capital and evolution of the urban system in Japan. *Ekistics, 50,* 110-122.

Mollenkopf, J. (Ed.). (1983). *The contested city.* Princeton, NJ: Princeton University Press.

Mollenkopf, J. (1993a). Cities in the new global economy. *American Prospect, 10*(13), 132-141.

Mollenkopf, J. (Ed.). (1993b). *Key urban nodes in the global system.* New York: Social Science Research Council.

Mollenkopf, J., & Castells, M. (Eds.). (1988). *Dual city: The restructuring of New York.* New York: Russell Sage.

Monroe, K. (1992). Tomorrow's cities are global cities. *American City & County, 107*(10), 42-44.

Moss, M. L. (1987a, September). *The informational city in the global economy.* Paper presented at the Third International Workshop on Innovation, Technology Change and Spatial Impacts, Cambridge, UK.

Moss, M. L. (1987b). Telecommunications, world cities and urban policy. *Urban Studies, 24*(6), 534-546.

Murakami, Y., & Kosai, Y. (Eds.). (1986). *Japan in the global community: Its role and contribution on the eve of the 21st century.* Tokyo: University of Tokyo Press.

Noyelle, T. (1986). *New York and the emergence of global financial markets.* New York: Regional Plan Association.

Noyelle, T. (1987, June). *Economic development: A look forward to the 1990's.* Paper presented at the Conference on the Future of the Metropolitan Economy: New Challenges for Policy and Policy-Oriented Research, TNO Research Center for Urban and Regional Planning, The Hague, The Netherlands.

Noyelle, T., & Stanback, T. M. (1985). *The economic transformation of American cities.* Totowa, NJ: Rowman & Allanheld.

Nusbaumer, J. (Ed.). (1987). *Services in the global market.* Dordrecht, The Netherlands: Kluwer.

O'Connor, K. (1991, September). *Economic activity in Australian cities: National and local trends and policy.* Paper presented at the conference on Productive Cities in the Twenty-First Century, Sydney, Australia.

O'Loughlin, J. (1992). *Between Stuttgart and Sheffield: Amsterdam in an integrated Europe and a competitive world economy.* Amsterdam: Centrum voor Grootstedelijk Onderzoek.

Osborne, D., & Gaebler, T. (1992). *Reinventing government.* Reading, MA: Addison-Wesley.

Pammer, W. J. (1992). The future of municipal finance in an era of fiscal austerity and economic globalization. In International City Managers Association (Ed.), *The municipal year book* (pp. 3-11). Washington, DC: International City Managers Association.

Panitch, L. (1994). Globalization and the state. In R. Milibrand & L. Panitch (Eds.), *Between globalism and nationalism.* New York: Monthly Review Press.

Parkinson, M., Bianchini, F., Dawson, J., Evans, R., & Harding, A. (1992). *Urbanisation and the functions of cities in the European Community.* Liverpool, UK: John Moores University, European Institute of Urban Affairs.

Paul, A. C. (Ed.). (1990). *Managing for tomorrow: Global change and local futures.* Washington, DC: International City Management Association.

Pickvance, C., & Preteceille, E. (Eds.). (1991). *State restructuring and local power: A comparative perspective.* London: Pinter.

Preteceille, E. (1990). Political paradoxes of urban restructuring: Globalization of the economy and localization of politics? In J. R. Logan & T. Swanstrom (Eds.), *Beyond city limits: Urban policy and economic restructuring in comparative perspective* (pp. 27-59). Philadelphia: Temple University Press.

Prud'homme, R. (1989). New trends in cities of the world. In R. V. Knight & G. Gappert (Eds.), *Cities in a global society* (Urban Affairs Annual Review, Vol. 35, pp. 44-57). Newbury Park, CA: Sage.

Pryke, M. (1991). An international city going global. *Environment and Planning D: Society and Space, 9,* 197-222.

Pugh, C. (1994). Housing policy development in developing countries: The World Bank and internationalisation. *Cities, 11*(3), 159-180.

Reynolds, J. I. (1988). Vancouver as international financial centre. *The Advocate, 46*(1), 95-98.

Rimmer, P. J. (1986). Japan's world cities: Tokyo, Osaka, Nagoya or Tokaido Megalopolis? *Development and Change, 17*(1), 121-158.

Rimmer, P. J. (1989). Urban change and the international economy. In M. P. Smith (Ed.), *Pacific Rim cities in the world economy* (Comparative and Community Research Series, Vol. 2, pp. 156-99). New Brunswick, NJ: Transaction.

Robertson, R. (1985). Modernization, globalization and the problem of culture in world-systems theory. *Theory, Culture & Society, 2*(3), 103-118.

Robertson, R. (1987). Globalization theory and civilization analysis. *Comparative Civilizations Review, 17,* 20-30.

Rodriguez, N. P., & Feagin, J. R. (1986). Urban specialization in the world system. *Urban Affairs Quarterly, 22*(2), 187-220.

Ross, A. (1991). Is global culture warming up? *Social Text, 28,* 3-30.

Ross, R., & Telkamp, G. (Eds.). (1985). *Colonial cities.* Dordrecht, The Netherlands: Martinus Nijhoff for the Leiden University Press.

Ross, R., & Trachte, D. (1983). Global cities and global classes: The peripheralization of labor in New York City. *Review, 6*(3), 393-431.

Ross, R., & Trachte, D. (1990). *Global capitalism: The new leviathan.* New York: State University of New York Press.

Rothblatt, D. (1992). San Jose: Emerging capital of Silicon Valley. In E. J. Blakely & R. J. Stimson (Eds.), *New cities of the Pacific Rim* (Monograph 43, pp. 4.1-4.25). Berkeley: University of California, Institute of Urban and Regional Development.

Sadler, D. (1992). *The global region: Production, state policies and uneven development.* New York: Pergamon.

Sassen, S. (1987). Growth and informalization at the core: A preliminary report on New York City. In M. P. Smith & J. R. Feagin (Eds.), *The capitalist city: Global restructuring and community politics* (pp. 138-154). Oxford, UK: Basil Blackwell.

Sassen, S. (1988). *The mobility of labour and capital: A study in international investment and labor flow.* Cambridge, UK: Cambridge University Press.

Sassen, S. (1990a). Beyond the city limits: A commentary. In J. R. Logan & T. Swanstrom (Eds.), *Beyond city limits: Urban policy and economic restructuring in comparative perspective* (pp. 237-242). Philadelphia: Temple University Press.

Sassen, S. (1990b). *The global city: New York, London, Tokyo.* Princeton, NJ: Princeton University Press.

Sassen, S. (1990c). *The interdependence of cities in the Committee on Urban Studies Report for 1989.* New York: UNESCO Committee on Urban Studies.

Sassen, S. (1994a). *Cities in a world economy.* London: Pine Forge.

Sassen, S. (1994b). The urban complex in a world economy. *International Social Science Journal, 139,* 43-62.

Sassen, S. (1994c). Urban impacts of economic globalization. In E. Isin (Ed.), *Toronto: Region in the world economy: A symposium on how the Toronto region is affected by changes in the world economy* (pp. 147-166). North York, ON: York University, Urban Studies Programme.

Sassen-Koob, S. (1984). The new labor demand in global cities. In M. P. Smith (Ed.), *Cities in transformation* (pp. 139-171). Beverly Hills, CA: Sage.

Savitch, H. V. (1987). *Post-industrial cities.* Princeton, NJ: Princeton University Press.

Savitch, H. V. (1990). Postindustrialism with a difference: Global capitalism in world-class cities. In J. R. Logan & T. Swanstrom (Eds.), *Beyond the city limits: Urban policy and the economic restructuring in comparative perspective* (pp. 150-175). Philadelphia: Temple University Press.

Savitch, H. V., & Kantor, P. (1994). Urban mobilization of private capital: A cross national comparison. Washington, DC: Woodrow Wilson International Center for Scholars.

Scanlon, R. (1989). New York as global capital in the 1980's. In R. V. Knight & G. Gappert (Eds.), *Cities in a global society* (Urban Affairs Annual Review, Vol. 35, pp. 83-95). Newbury Park, CA: Sage.

Sclar, E. D., & Hook, W. (1993). The importance of cities to the national economy. In H. G. Cisneros (Ed.), *Interwoven destinies: Cities and the nation* (pp. 48-80). New York: Norton.

Seiko, S. (1987). *Internationalisation and regional structure.* Chiba, Japan: Chiba University.

Shachar, A. (1994). Holland: A world city? *Urban Studies, 31*(3), 381-400.

Shefter, M. (Ed.). (1993). *Capital of the American century: The national and international influence of New York City.* New York: Russell Sage.

Shuman, M. H. (1987). *Building municipal foreign policies* (Vol. 86). Irvine, CA: Center for Innovative Diplomacy.

Shuman, M. H. (1992). Courts v. local foreign policies. *Foreign Policy, 86,* 158-177.

Shuman, M. H. (1994). *Towards a global village: International community development initiatives.* London/Boulder, CO: Pluto Press in association with the Institute for Policy Studies and Towns and Development.

Simon, D. (1989). Colonial cities, post colonial Africa and the world-economy: Some pertinent issues and questions. *International Journal of Urban and Regional Research, 13*(1), 28-43.

Slater, D. (1986). Capitalism and urbanisation at the periphery. In D. Drakis-Smith (Ed.), *Urbanisation in the developing world* (pp. 94-112). London: Croom Helm.

Smidt, M. de. (1992). A world-city paradox: Firms and the urban fabric. In F. M. Deileman & S. Musterd (Eds.), *The Ranstad: A research and policy laboratory* (pp. 97-122). Dordrecht, The Netherlands: Kluwer.

Smith, C. A. (1985). Theories and measures of urban primacy: A critique. In M. Timberlake (Ed.), *Urbanization in the world economy* (pp. 87-117). Orlando, FL: Academic Press.

Smith, D., & London, B. (1990). Convergence in world urbanization: A quantitative assessment. *Urban Affairs Quarterly, 25,* 574-590.

Smith, D., & Timberlake, M. (1993, April). *Mapping world-system city networks: Conceptual and operational difficulties.* Paper presented at the conference on World Cities in a World System, Sterling, VA.

Smith, M. P. (1987). Global capital restructuring and local political crisis in U.S. cities. In J. Henderson & M. Castells (Eds.), *Global restructuring and territorial development* (pp. 234-250). Newbury Park, CA: Sage.

Smith, M. P. (1988). *City, state and market: The political economy of urban society.* New York: Basil Blackwell.

Smith, M. P. (Ed.). (1984). *Cities in transformation.* Beverly Hills, CA: Sage.

Smith, M. P. (Ed.). (1989). *Pacific Rim cities in the world economy* (Comparative and Community Research Series, Vol. 2). New Brunswick, NJ: Transaction.

Smith, M. P., & Feagin, J. R. (Eds.). (1987). *The capitalist city: Global restructuring and community politics.* Oxford, UK: Basil Blackwell.

Smith, P. J., & Cohn, T. H. (1994). International cities and municipal paradiplomacy: A typology for assessing the changing Vancouver metropolis. In F. Frisken (Ed.), *The changing Canadian metropolis: A public policy perspective* (Vol. 2, pp. 613-655). Berkeley: University of California, Institute of Governmental Studies; Toronto, ON: Canadian Urban Institute.

Soja, E. W. (1987). Economic restructuring and the internationalisation of the Los Angeles region. In M. P. Smith & J. R. Feagin (Eds.), *The capitalist city: Global restructuring and community politics* (pp. 178-198). Oxford, UK: Basil Blackwell.

Stevenson, D. (1994). The new economy, the new cities: Retrospect. In E. Isin (Ed.), *Toronto: Region in the world economy: A symposium on how the Toronto region is affected by changes in the world economy* (pp. 223-229). North York, ON: York University, Urban Studies Programme.

Stimson, R. J. (1993) *Process of globalization and economic restructuring and the emergence of a new space economy of cities and regions in Australia.* Paper presented at the Fourth International Workshop on Technological Change and Urban Form: Productive and Sustainable Cities, Berkeley, CA.

Stohr, W. B. (1990). *Global challenge and local response: Initiatives for economic regeneration in contemporary Europe.* Lawrence: University Press of Kansas.

Stoker, G. (1988). *The politics of local government.* Hound Mills, UK: Macmillan Education.

Stone, C. N. (1991). The hedgehog, the fox and the new urban politics: Rejoinder to Kevin R. Cox. *Journal of Urban Affairs, 13,* 289-298.

Stone, C. N., & Sanders, H. W. (Eds.). (1987). *The politics of urban development.* Lawrence: University Press of Kansas.

Storper, M. (1991). *Industrialisation, economic development, and the regional question in the third world.* London: Pion.

Storper, M. (1992). The limits to globalization: Technology districts and international trade. *Economic Geography, 68,* 60-93.

Stutz, F. (1992). San Diego: The next high-amenity Pacific Rim world city. In E. J. Blakely & R. J. Stimson (Eds.), *New cities of the Pacific Rim* (Monograph 43, pp. 5.1-5.26). Berkeley: University of California, Institute of Urban and Regional Development.

Swanstrom, T. (1986). Urban populism, fiscal crisis and the new political economy. In M. Gottdiener (Ed.), *Cities in stress: A new look at the urban crisis* (Urban Affairs Annual Review, Vol. 30, pp. 81-110). Beverly Hills, CA: Sage.

Swanstrom, T. (1994). Globalization: A critical view. In E. Isin (Ed.), *Toronto: Region in the world economy: A symposium on how the Toronto region is affected by changes in the world economy* (pp. 3-14). North York, ON: York University, Urban Studies Programme.

Tardanico, R., & Larin, R. M. (Eds.). (in press). *Reestructuración transnacional, empleo urbano y desigualdad social: Cambio comparativo en Costa Rica, Guatemala, Mexico y Venezuela* [Transnational restructuration, urban employment and social disequality: Comparative change in Costa Rica, Guatemala, Mexico and Venezuela]. San José, Costa Rica: Facultad Latinoamericana de Ciencias Sociales (FLACSO).

Tardanico, R., & Lungo, M. (in press). Local dimensions of local restructuring: Changing labour-market contours in urban Costa Rica. *International Journal of Urban and Regional Research.*

Terasaka, A. et al. (1988). The transformation of regional systems in an information-oriented society. *Geographical Review of Japan, 61*(1), 159-173.

Thrift, N. (1987). The fixers: The urban geography of international commercial capital. In J. Henderson & M. Castells (Eds.), *Global restructuring and territorial development* (pp. 203-233). Newbury Park, CA: Sage.

Thrift, N. (1992). Muddling through: World orders and globalization. *The Professional Geographer, 44*(2), 3-7.

Timberlake, M. (1987). World-system theory and the study of comparative urbanisation. In M. P. Smith & J. R. Feagin (Eds.), *The capitalist city: Global restructuring and community politics* (pp. 37-65). Oxford, UK: Basil Blackwell.

Timberlake, M. (Ed.). (1985). *Urbanization in the world-economy.* Orlando, FL: Academic Press.

Todd, G. (1993). *The political economy of urban and regional restructuring in Canada: Toronto, Montreal and Vancouver in a global cconomy 1970-90.* Unpublished doctoral dissertation, York University, Department of Political Science, North York, ON.

Trachte, K., & Ross, R. (1985). The crisis of Detroit and the emergence of global capitalism. *International Journal of Urban and Regional Research, 9,* 216-217.

Vogel, D. (1988). *The future of New York City as a global and national financial center.* Paper presented to the Metropolitan Dominance Working Group, Social Science Research Council Committee on New York City, New York.

Walters, P. B. (1985). Systems of cities and urban primacy: Problems of definition and measurement. In M. Timberlake (Ed.), *Urbanization in the world economy* (pp. 63-85). Orlando, FL: Academic Press.

Walton, J. (1982). The international economy and peripheral urbanization. In N. I. Fainstein & S. S. Fainstein (Eds.), *Urban policy under capitalism* (Urban Affairs Annual Review, Vol. 22). Beverly Hills, CA: Sage.

Ward, K. B. (1985). Women and urbanization in the world system. In M. Timberlake (Ed.), *Urbanization in the world economy* (pp. 305-323). Orlando, FL: Academic Press.

Wheeler, J. O. (1988). The economic role of large metropolitan areas in the United States. *Growth and Change, 19,* 75-86.

Wilmoth, D. (1992). Tiajin's urban development and the open door. In E. J. Blakely & R. J. Stimson (Eds.), *New cities of the Pacific Rim* (Monograph 43, pp. 17.1-17.23). Berkeley: University of California, Institute of Urban and Regional Development.

Wu, C. T. (1992). Taipei: From industrial center to international city. In E. J. Blakely & R. J. Stimson (Eds.), *New cities of the Pacific Rim* (Monograph 43, pp. 12.1-12.33). Berkeley: University of California, Institute of Urban and Regional Development.

Wunsch, J. S., & Olowu, D. (Eds.). (1990). *The failure of the centralized state.* Boulder, CO: Westview.

Wusten, H. van der. (1992). A new world order. *The Professional Geographer, 44*(2), 19-22.

Zelinsky, W. (1991). The twinning of the world: Sister cities in geographic and historical perspective. *Annals of the Association of American Geographers, 81*(1), 1-31.

Zukin, S. (1982). *Loft living, culture and capital in urban change.* Baltimore: Johns Hopkins University Press.

Zukin, S. (1992). The city as a landscape of power: London and New York as global financial capitals. In L. Budd & S. Whimster (Eds.), *Global finance and urban living: A study of metropolitan change* (pp. 195-223). London: Routledge.

Index

About the Editors

Peter Karl Kresl is Professor of Economics at Bucknell University. Educated at the University of Texas, he is a specialist in international economics, Canada-United States economic relations, the economics of Western European integration, and cultural economics. During the past 5 years, the primary focus of his research has been urban economics and their relation to the international economy. His book *The Urban Economy and Regional Trade Liberalization* was published in 1992. In 1994, he completed a research project on "The Competitiveness of Cities: The United States" for the Organization for Economic Cooperation and Development, and he has published papers on urban economies in recent issues of *Ekistics, The Journal of European Integration,* and *Quebec Studies.* He has participated in the New International Cities Era research group and serves on the advisory panel for the United Nations Conference Habitat II. He has been guest professor in Canada, Norway, and Sweden and has lectured often in Europe, North America, and Asia.

Gary Gappert is Director of the Institute for Futures Studies and Reseach at the University of Akron where he also serves as Professor of Public Administration and Urban Studies. His current work includes publications on urban environmental policies and issues, a major conference in 1997-1998 on Machiavelli, urban leaders and cities in the 21st century, and a book on "vision and method" in future studies. His recent international work includes public lectures in eastern Europe and consulting for the Peace Corps in Africa on urban planning projects. With the publication of this volume, Professor Gappert has reached his goal of editing four *Urban Affairs Annuals.* His previous volumes include *The Social Economy of Cities* (with Harold Rose), *The Future of Winter Cities,* and *Cities in the Global Society* (with Richard V. Knight). At Akron, he is currently developing an urban millenium project for which he is inviting participation.

About the Contributors

Alan F. J. Artibise is Professor and former Director of the School of Community and Regional Planning at the University of British Columbia in Vancouver. Between 1993 and 1995, he also served as Executive Director of the International Centre for Sustainable Cities in Vancouver. He is actively involved in the Cascadia Region and has lectured and written extensively on this emerging transborder city region.

Norris C. Clement is Professor of Economics at San Diego State University, San Diego, CA. His recent research focuses on the development of transborder regions throughout the world. Currently, he is codirector of a project to develop an integrated, interregional model of the Southern California-Baja California region.

Theodore H. Cohn is Professor of Political Science at Simon Fraser University, Burnaby, British Columbia, and served as Chair of the Political Science Department from 1982 to 1987. He is the author of numerous publications on agricultural trade policy, including *The International Politics of Agricultural Trade: Canadian-American Relations in a Global Agricultural Context,* and has also written extensively on agricultural trade negotiations in the NAFTA and GATT/World Trade Organization.

Earl H. Fry is Professor of Political Science and Endowed Professor of Canadian Studies at Brigham Young University, Provo, UT. He is the author of *America the Vincible: U.S. Foreign Policy for the Twenty-first Century* (1994) and editor of *States and Provinces in the International Economy* (1993) and *The New International Cities Era* (1989). During the 1995-1996 academic year, he is serving as the Bissell-Hyde Fulbright Professor at the University of Toronto.

Paul Gregory is Information Specialist at the Centre for Urban and Community Studies at the University of Toronto. A graduate of the University of Toronto Faculty of Information Studies in 1992, he has held positions at the Municipality of Metropolitan Toronto, the City of Toronto, and Government Publications at the University of Toronto. He is currently researching information systems management at the urban level.

Daniel Hiernaux was educated in regional and urban science at Leuvian University in Belgium (M.S.) and in Latin American Studies at the Sorbonne (Ph.D.). He worked for the government of Mexico for 10 years and in 1984 joined the faculty of the Metropolitan Autonomous University in Mexico City. He is author of seven books and of 150 scholarly papers, in French, English, Spanish, and Portuguese. He has also had training in urban affairs at the World Bank and the United Nations.

John Kincaid is the Robert B. and Helen S. Meyner Professor of Government and Public Service and Director of the Meyner Center for the Study of State and Local Government at Lafayette College, Easton, PA. He served as Executive Director of the U.S. Advisory Commission on Intergovernmental Relations, Washington, DC, from 1988 to 1994. He is Editor of *Publius: The Journal of Federalism,* has written extensively on federalism and intergovernmental relations, and coedited *Competition Among States and Local Governments: Equity and Efficiency in American Federalism* (1991).

Marc V. Levine is Associate Professor of History and Urban Studies at the University of Wisconsin—Milwaukee, where he is Director of Urban Studies Programs and Director of the Center for Economic Development. He is the author of *The Reconquest of Montreal: Language Policy and Social Conflict in a Bilingual City,* and coauthor of *The State and Democracy: Revitalizing America's Government,* as well as numerous articles in academic journals and anthologies.

Pierre-Paul Proulx is Professeur titulaire, Département de Sciences économiques, Université de Montréal, and Vice-président, Conseil des relations internationales de Montréal. He is also former Assistant Deputy Minister with the federal government in Ottawa. His fields of interest include trade policy, economic integration, international cities, and economic development policies.

Samuel Schmidt is Associate Professor of Political Science and Director of the Center for Inter-American and Border Studies at the University of Texas, El Paso, and is also a Research Associate in the UCLA Program on Mexico. His books include *The Deterioration of the Mexican Presidency* and *La Autonomia Relativa del Estado* (The Relative Autonomy of the State). He is the author (with David Lorey) of *Policy Recommendations for Managing the El Paso-Ciudad Juárez Metropolitan Area.* He is the editor of *Enfrentando el Futuro* (Facing the Future) (1990) and the *United States-Mexico Border Environmental Directory* (1994).

Arie Shachar is Professor of Urban Geography and Planning and Director of the Institute of Urban and Regional Studies at the Hebrew University of Jerusalem, Israel. He specializes in metropolitan development and planning, economic restructuring and urban dynamics, and issues of European integration and its impact on urban change. He is codirector of the European Science Foundation research program on Regional and Urban Restructuring in Europe and serves as a consultant to various internaional organizations dealing with economic development and urban policy formation.

Patrick J. Smith is Associate Professor of Political Science at Simon Fraser University in British Columbia. He is the author or coauthor of three books on Canadian politics and has published articles in periodicals including the *Canadian Journal of Political Science, Planning, and Administration* and the *Canadian Journal of Urban Research.* He has been an adviser to the government of British Columbia and continues his research focus on metropolitan governance and on the Greater Vancouver urban region.

Wilbur R. Thompson is Professor Emeritus in Economics at Wayne State University, Detroit, MI. He is the author of *A Preface to Urban Economics* and numerous articles on regional and urban economics. From 1985 to 1986, he held the Albert A. Leven Chair of Urban Studies and Public Service at Cleveland State University. He received his Ph.D. in economics from the University of Michigan (1953). He now resides in Albuquerque, NM.